Fillers and Reinforcements for Plastics

Rudolph D. Deanin and
Nick R. Schott, *Editors*

A symposium co-sponsored by the ACS Division of Organic Coatings and Plastics Chemistry and the Plastics Institute of America at the 166th Meeting of the American Chemical Society, Chicago, Ill., Aug. 27–29, 1973.

ADVANCES IN CHEMISTRY SERIES **134**

AMERICAN CHEMICAL SOCIETY
WASHINGTON, D. C. 1974

Library of Congress CIP Data

Fillers and reinforcements for plastics
(Advances in chemistry series, 134)

Includes bibliographical references.

1. Reinforced plastics—Congresses.
I. Deanin, Rudolph D., 1921- ed. II. Schott, Nick R., 1939- ed. III. American Chemical Society. Division of Organic Coatings and Plastics Chemistry. IV. Plastics Institute of America, Hoboken, N. J. V. Series.

QD1.A355 no. 134 (TP1177) 540'.8s 668.4'94
74-13562
ISBN 0-8412-0202-8 ADCSAJ 134 1-224 (1974)

PRINTED IN THE UNITED STATES OF AMERICA

Advances in Chemistry Series

Robert F. Gould, *Editor*

FOREWORD

ADVANCES IN CHEMISTRY SERIES was founded in 1949 by the American Chemical Society as an outlet for symposia and collections of data in special areas of topical interest that could not be accommodated in the Society's journals. It provides a medium for symposia that would otherwise be fragmented, their papers distributed among several journals or not published at all. Papers are refereed critically according to ACS editorial standards and receive the careful attention and processing characteristic of ACS publications. Papers published in ADVANCES IN CHEMISTRY SERIES are original contributions not published elsewhere in whole or major part and include reports of research as well as reviews since symposia may embrace both types of presentation.

CONTENTS

Preface vii

1. **The Role of Fillers and Reinforcements in Plastics Chemistry** 1
   Raymond B. Seymour

2. **The Coatings Industry Looks at Fillers** 7
   Fred B. Stieg

3. **Asbestos as a Reinforcement and Filler in Plastics** 16
   John W. Axelson

4. **Asbestos-Reinforced Rigid Poly(vinyl chloride)** 29
   A. Crugnola and A. M. Litman

5. **Mica as a Reinforcement for Plastics** 41
   P. D. Shepherd, F. J. Golemba, and F. W. Maine

6. **Low Quartz Microforms and Their Contributions to Plastics and Resinous Systems** 52
   James E. Moreland

7. **Acrylic Modification of Plasticized Poly(vinyl chloride)** 61
   John T. Lutz, Jr.

8. **Silane Coupling Agent for Reinforcing Mineral-Filled Nylon** 73
   S. E. Berger, P. J. Orenski, and M. W. Ranney

9. **Catalytic Effects in Bonding Thermosetting Resins to Silane-Treated Fillers** 86
   Edwin P. Plueddemann

10. **The Rheology of Concentrated Suspensions of Fibers. I. Review of the Literature** 95
    Richard O. Maschmeyer and Christopher T. Hill

11. **Compounding of Fillers in Motionless Mixers** 106
    Nick R. Schott, Stephen A. Orroth, and Arunkumar Patel

12. **Compounding of Fillers** 114
    Stan Jakopin

13. **Filler Reinforcement of Plasticized Poly(vinyl chloride)** 128
    Rudolph D. Deanin, Raymond O. Normandin, and Ghanshyam J. Patel

14. **The Effects of Moisture on the Properties of High Performance Structural Resins and Composites** 137
C. E. Browning and J. M. Whitney

15. **Properties of Filled Polyphenylene Sulfide Compositions** 149
H. Wayne Hill, Jr., Robert T. Werkman, and G. E. Carrow

16. **Biodegradable Fillers in Thermoplastics** 159
Gerald J. L. Griffin

17. **Performance of Conductive Carbon Blacks in a Typical Plastics System** 171
J. H. Smuckler and P. M. Finnerty

18. **Hydrated Alumina as a Fire-Retardant Filler in Styrene–Polyester Casting Compounds** 184
C. V. Lundberg

19. **Corrosion Engineering in Reinforced Plastics** 195
Otto H. Fenner

20. **Reinforced Plastics in Low Cost Housing** 207
Armand G. Winfield and Barbara L. Winfield

Index 219

PREFACE

Polymer synthesis and basic theory are concerned primarily with pure polymer systems. In commercial practice, however, most polymers are mixed with other solid materials to improve their balance of properties. Early use of carbon black in rubber improved most properties and was therefore referred to as "reinforcement." The use of fibers in plastics also improves many properties dramatically, forming the basis for "reinforced plastics." In addition to fibers, other types of solid fillers are used to provide a great variety of beneficial effects, some of them quite complex and subtle.

The use of fillers and reinforcements in plastics thus rests on a great deal of practical commercial experience, which has been followed by much theoretical analysis and advanced exploratory research. Scientific discussion has concentrated most frequently among plastics engineers, with some assistance from polymer chemists. To enhance and stimulate the contributions of polymer and plastics chemists to this field, a symposium on Fillers and Reinforcements for Plastics was organized as part of the Organic Coatings and Plastics Chemistry Division program at the American Chemical Society national meeting in Chicago on August 27-29, 1973. The papers presented at that symposium have been expanded and revised for publication in this volume.

It is hoped that this collection will prove interesting and stimulating to the many polymer and plastics chemists and engineers working in the field of fillers and reinforcements. In the words of Raymond B. Seymour, our lead speaker at the symposium, we should reassemble the symposium in a few years, view the progress we have made in the interim, and plan to launch our next steps into the future.

RUDOLPH D. DEANIN and NICK R. SCHOTT

Lowell, Mass.
February 1974

1

The Role of Fillers and Reinforcements in Plastics Chemistry

RAYMOND B. SEYMOUR

University of Houston, Houston, Tex. 77004

As with natural composites such as wood, leaves, feathers, and bones, the resin-like component in modern synthetic composites transfers the stress to the reinforcing component. Thus a strong interfacial bond is essential, and crack propagation must be deterred in high strength composites. The discontinuous phase may consist of silicas, silicates, calcium carbonate, carbon black, and comminuted polymers or more functional reinforcements such as fibrous glass, graphite, boron and single crystals. While the reinforcements have been used primarily with thermosetting resins in the past, they are now being used to reinforce thermoplastics, and these new composites have added a new dimension to the plastics industry.

Most modern and traditional organic coatings and plastics contain additives, and hence are composites in a broad sense. Additives include gases in cellular plastics, pigments in coatings, plasticizers in flexibilized poly(vinyl chloride), antioxidants in weather-resistant plastics, lubricants, and gloss-control agents as well as fillers and reinforcements. This discussion is limited to composites containing fillers and reinforcements only.

Although such a discussion could include materials such as wood, leaves, feathers, and bones, it is confined to composites consisting of a synthetic resinous or continuous phase and a natural occurring or synthetic discontinuous or filler phase. The strength of these composites depends on the interfacial bond between these components, the ability of the composite to deter crack propagation, the strength of the discontinuous phase, and, to a lesser extent, the strength of the continuous resinous phase.

When there is little interfacial action, as is the case with glass sphere–resin composites, the modulus of the composite (M) is related to the modulus of the resin (M_o), and the fractional volume occupied by the filler (c) as shown by the following Einstein-Guth-Gold (EGG) equation:

$$M = M_o(1 + 2.5c + 14.1c^2)$$

Since there is some interfacial bonding between the filler surface and the resin and the particles tend to cluster, the EGG equation must be modified for carbon black–resin composites as follows:

$$M = M_o(1 + 0.67fc + 1.62f^2c^2)$$

The shape factor (f) is equal to the ratio of the length to the diameter of the filler particles. The ratio of the diameter—*i.e.*, thickness—is called the aspect ratio for lamellar fillers.

When c is greater than 0.1, it is essential to include a crowding factor, β, as suggested by Mooney. The Mooney equation for simple inert spheres is:

$$M = M_o\left(\frac{2.5c}{1 - \beta c}\right)$$

The use of silicas, such as fumed silica and quartz and silicates, such as clay and talc which are widely used as fillers, is discussed in other chapters in this volume. These fillers may be spheroidal, lamellar, or acicular. Surface treatment which increases the strength of the interfacial bond is also discussed in other chapters as are relevant applications of composites on corrosion engineering and housing.

The rubber industry and the plastics composite industry were built primarily on empirical knowledge. However, the future growth of these composite industries will depend largely on the intelligent use of performance data and modern concepts of the interfacial bond between the continuous and discontinuous phases of composites. The latter is sometimes called micromechanics.

The history of the composite industry includes numerous successes as well as some failures. Failures for composites, such as reinforced plastic pipe have usually resulted from an interfacial bond failure. This type of failure is aggravated when transitions occur as a result of changes in temperature during service. Thus, failures may occur when amorphous plastic composites are heated above their glass transition temperatures, when crystalline plastic composites are heated above their melting points, and when thermosetting resin composites are heated above their useful temperatures.

As proof of their numerous successful applications, over 1.2 billion pounds of fibrous glass reinforced plastics were used in the United States in 1972 (*1*). Because of their outstanding performance in marine, transportation, construction, and corrosion-resistant applications, it is expected that the consumption of composites will double within the next five years. Annual past and predicted consumption data for reinforced plastics are listed in Table I.

Table I. Annual Consumption in Millions of Pounds of Reinforced Plastics in the United States

Application	*1967*	*1972*	*1978*
Marine	127	321	550
Transportation	80	252	700
Construction	73	185	450
Consumer	57	99	115
Electrical	45	79	85
Corrosion resistant	40	110	390
Aircraft	24	28	40
Appliances	18	61	180
Miscellaneous	90	93	180
Total	544	1,227	2,800

Because of their high strength-to-weight ratio, durability, and freedom of design, the worldwide growth of these composites should be comparable with the growth data shown for the United States in Table I. Certainly, the use of composite materials in Western Europe should double by 1979 (*2*). The data on the consumption of nonfibrous fillers are not as readily delineated as those on fibrous reinforcements. The use of filled plastics has been reviewed annually for many years (*3, 4, 5*).

Wood flour, which is made by attrition grinding of wood, is superior as a filler to the less fibrous ground shell flour. Wood flour was used by Baekeland at the beginning of the 20th century and is still used to reinforce phenolic resins. Composites with higher impact resistance are obtained by replacing the wood flour by cellulosic, asbestos, glass, and nylon fibers. The art which was developed for phenolic resin composites has been extended to the reinforcement of polyester and epoxy thermosetting resins.

Fibrous glass reinforced polyester composites now account for over 80% of the volume of plastic composites. This phase of the industry started with composites made by the hand laying up of resin impregnated glass mat. However, much of the recent growth has been associated with bulk molding compounds (BMC) and sheet molding compounds (SMC).

Injection molded reinforced polyester resin composites have also contributed to the growth of the reinforced segment of the plastics industry (*6*).

Despite limited production, the role of reinforcements in high performance composites, such as reinforced polyimides, is also impressive (*7*). The superiority of the latter as structural materials has justified the use of more sophisticated reinforcements, such as sapphire single crystals and graphite and boron filaments. The high cost of sapphire reinforcements has been reduced by spinning these products from molten alpha alumina.

Reinforcement was considered to be limited to thermosetting composites prior to 1950. This art has now been extended to thermoplastics, and one of the early applications was the production of fibrous glass-reinforced polystyrene minesweepers in 1955. However, crystalline thermoplastics like nylon, rather than amorphous thermoplastics like polystyrene, are preferred. A process for making fibrous glass-reinforced nylon molding powders was patented in 1958 (*8*).

Table II. Annual Consumption in Millions of Pounds of Reinforced Thermoplastics in 1971 and 1972

Type of Plastic	*1971*	*1972*
Polypropylene	22	28
Styrene polymers	19	20
Nylon	12	16
Polyethylene	6	7
Acetal	3	4
Miscellaneous	5	7
Total	67	82

Reinforced thermoplastics (RTP) are engineering plastics which perform satisfactorily under conditions where unfilled thermoplastics fail. As shown in Table II, the annual production of RTP in the United States, was less than one million pounds in 1964, grew to more than 80 million pounds or almost 40,000 metric tons in 1972 (*9*). The current annual consumption of RTP in Western Europe is 20,000 metric tons. Nylon accounts for more than half of this volume. Consumption of RTP in Europe is expected to triple within the next five years (*10*).

Potassium titanate microfibers are being used to produce composites of nylon, acetal, polypropylene, and ABS. Because they are small, these single crystal mineral reinforcements are more randomly oriented than the larger fibrous glass particles. Thus, moldings of these composites are essentially isotropic whereas fibrous glass–nylon composites are anisotropic. This sophisticated filler has also been used to produce ABS composites that can be electroplated (*11*).

As shown in Figure 1, nylon composites are now being used successfully as pump and compressor parts. These parts operate at high temperatures and pressures and outperform previously used metal parts.

The molded parts shown contain fibrous glass which increases the rigidity, tensile strength, creep resistance, and temperature resistance and reduces the thermal coefficient of expansion and moisture absorption of nylon 6,6. This composite also contains a polytetrafluoroethylene (PTFE) filler which lubricates the moving parts. Thus, the performance of these composites demonstrates the role of both fillers and reinforcements in plastics technology.

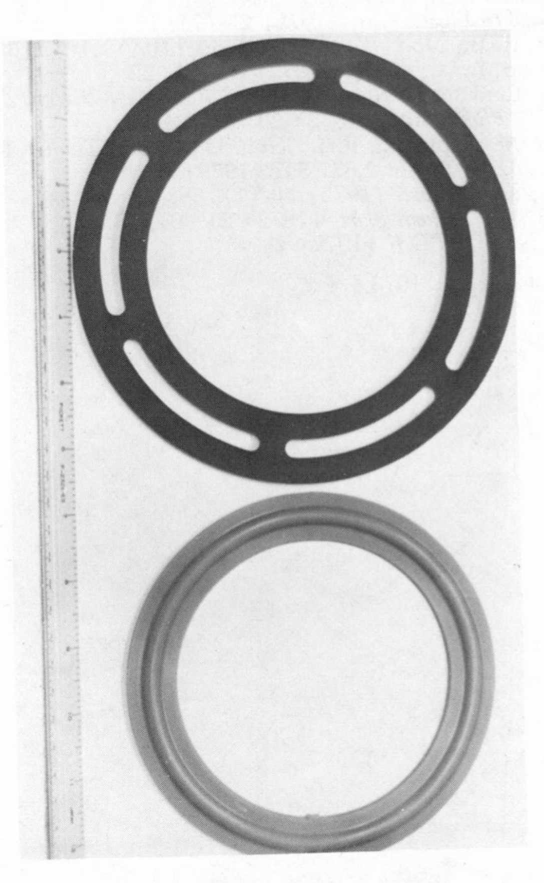

Figure 1. Molded fibrous glass parts

Most of the 24 billion pounds of plastics produced in the United States in 1972 was used for general purpose applications. While the more difficult problems were solved by sophisticated plastic composites, the more economical fibrous glass reinforced polyesters continued to account for the bulk of the consumption of reinforced plastics. One of the major factors in this growth was SMC.

The rate of future growth of RTP will be equally dramatic. Potassium titanate microcrystals and other economical whisker-like fillers will contribute to this growth. Leaders in all segments of the plastics industry are confident that plastic composites will continue to solve problems that cannot be solved with traditional materials of construction.

Literature Cited

1. *Chem. Week* (1973) **112** (8), 27.
2. Wildman, D., *Plast. Aust.* (1972) **23** (7), 15.
3. Seymour, R. B., *Mod. Plast.* (1973) **50** (10A) 210, 216.
4. Schoengood, A. A., *SPE J.* (1972) **28** (6), 22.
5. Seymour, R. B., *Ann. Rev. Ind. Eng. Chem., 1970* (1972) 305.
6. Austin, C., *Plast. Aust.* (1972) **23** (3), 13.
7. Witzel, J. M., Jablonski, R. J., Kruh, D., *Chem. Tech.* (1972) **2,** 440.
8. Bradt, R., U. S. Patent **2,877,501** (1959).
9. Gross, S., *Mod. Plast.* (1973) **50** (1), 59.
10. Wellman, J. F., *Plast. Aust.* (1972) **23** (6), 13.
11. Weston, N. E., *SPE J.* (1972) **28,** 37.

RECEIVED October 11, 1973.

2

The Coatings Industry Looks at Fillers

FRED B. STIEG

Titanium Pigment Division, N L Industries, Inc., South Amboy, N. J. 08879

Pigment packing characteristics have become an important formulating consideration in the paint industry. The binder demand of any given combination of pigments and/or fillers can be determined from its oil absorption, and it can be expressed as its critical pigment volume concentration (CPVC). A relationship exists between the CPVC of a pigmentation and the viscosity that it will produce in a given vehicle system:

$$\eta_R = \frac{P}{\mathrm{CPVC} - P}$$

where η_R *is the relative kinematic viscosity, and* P *is per cent pigment and/or filler by volume. Since fillers can be blended to maximize CPVC, it follows that maximum filler contents in polymer systems can be increased by a similar procedure.*

The materials known in the plastics industry as fillers are defined by ASTM's Committee D-20 as "relatively inert material added to a plastic to modify its strength, permanence, working properties, or other qualities, or to lower cost" (*1*). This definition includes, however, fibrous materials and fabrics, such as those used in plastic laminates, which primarily accounts for any reference to the modification of strength or permanence. The materials which are sometimes referred to in the coatings industry as fillers—but more commonly as extenders—are the nonfibrous, inert, inorganic pigments most frequently used in the plastics industry solely to reduce cost. Typically they include calcium carbonate (or whiting), barium sulfate, calcium sulfate, silica, china clay, magnesium silicate, powdered mica, diatomaceous silica, and so forth.

The common use of the term extender no doubt originated from the Webster's Dictionary interpretation, "to increase the bulk of a product by the addition of a cheaper substance : adulterate." In the early days of the coatings industry, fillers or extenders were branded as adulterants by legal statutes, and early labeling laws were designed to force paint manufacturers to disclose the use of any materials other than "pure" linseed oil, white lead, and zinc oxide in their products. Such early attempts by the federal government to regulate private industry specified the use of those same lead pigments that today's government regulations are designed to eliminate.

The coatings industry, however, has had many reasons to think of fillers as serving some purpose other than adulteration. As early as 1907, Perry (*2*) had drawn an analogy between paint and concrete (referring to a paper by Taylor and Thompson, Watertown Arsenal) and had devised a so-called "law of minimum voids" in which he concluded that "the requisite thickness of a paint film, together with the utmost attainable strength and impermeability, can best be obtained by a properly proportioned blend of pigments of three or more determinate sizes." Perry conceived of the largest particles in such a blend of pigments as piers, acting as physical supports for the dry paint film, which he further described as a series of flat arches. There is no record of any experimental work ever having been done to support these conclusions, but in the early 1900's an extensive series of coatings was exposed on the test fences of the North Dakota Experimental Station to establish the value of fillers in exterior house paints. The results appeared to support Perry's hypothesis in that many paint films containing coarse fillers proved to be more durable than those pigmented only with the so-called "pure" pigments.

A statistical analysis of the North Dakota data by Calbeck (*3*) showed that paint films containing less than 26% pigment by volume were significantly less durable than those containing 26–30% pigment by volume. Above this value, durability was again impaired.

These findings did not specify that any of this pigment volume need necessarily be composed of filler, but economic considerations dictated that any quantity of pigment above that required to produce acceptable brightness and opacity would be just that. Furthermore, the so-called pure pigments—white lead and zinc oxide—were chemically reactive with common paint vehicles, so that an excess had to be avoided to prevent paint failures such as alligatoring, checking, and cracking. Fillers, on the other hand, were inert and could inhibit such failures, if fibrous or platy, by apparently reinforcing the film against the mechanical stress from poor dimensional stability of wood substrates. In addition, fillers were necessary to produce lusterless, flat finishes. Their larger, more irregularly shaped particles were far more efficient in producing a light-diffusing surface than white lead or zinc oxide. In fact, such fillers were frequently called flattening agents.

Titanium Dioxide

With the appearance of a new, non-reactive white pigment, titanium dioxide, new emphasis was placed upon the use of fillers. Titanium pigments were so much more opaque than white lead or zinc oxide that substantital quantities of filler could be added. Not only could fillers be added, they were necessary for maximum efficiency in many types of formulations. This was caused by a peculiarity of white hiding pigments that had never been noticed before because of the low opacity of the older types. As the concentration of the hiding pigment was increased in a paint film, the amount of hiding power contributed per pound of pigment became smaller. This is illustrated in Figure 1 for rutile titanium dioxide. This effect is attributed to the fact that light scattered by an individual particle will encounter interference from light scattered, out-of-phase, by adjacent particles, and the more closely such particles approach each other, the more the interference, and the less total light scattered from the film to produce whiteness, brightness, and opacity.

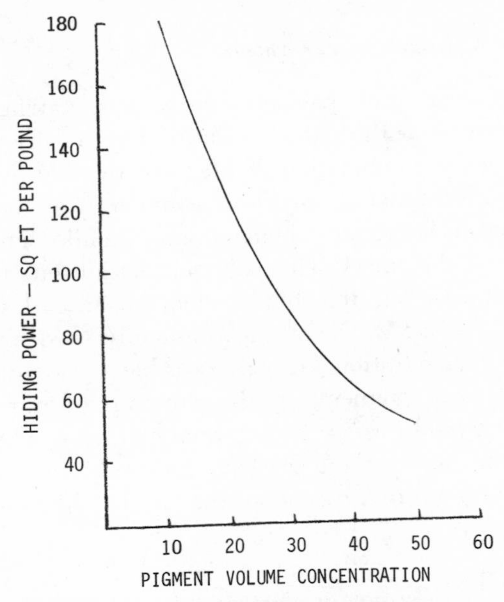

Figure 1. Hiding power curve for rutile TiO_2

The use of more than 30 vol % titanium dioxide in a paint film results in an actual loss of opacity because of this interference effect (*4*), yet many types of coatings must contain substantially more pigment than this to develop desired performance characteristics. Flat wall paints, for example, typically contain 55–65% pigment by volume (dry-film basis), and ceiling paints may contain more than 70%. If no more than 30% can be titanium dioxide, the balance must be filler. In practice, the use of more than 21% titanium dioxide by volume is uneconomical in terms of hiding power per dollar of total cost.

When large amounts of filler began to be used, interesting effects of particle size were noted. For example, the combination of titanium dioxide with the types of coarse filler commonly used in flat wall paints to produce a lustreless finish resulted in an unexpected loss of hiding power. This was the result of the packing phenomenon noted 50 years earlier by Perry. The fine titanium dioxide particles were being packed into the voids of the coarse extender system by the forces of film formation, and their proximity was exaggerating the interference effect. This undesirable condition could be alleviated by adding a portion of filler of average particle diameter of the same order of magnitude as that of titanium dioxide. Since it was similar in size, it found its way into the same voids and provided a beneficial physical dilution of the titanium dioxide.

Critical Pigment Volume Concentration

The evidence that such packing effects were taking place led to extensive investigation of the volume relationships in dry paint films. It was found that every combination of titanium dioxide and filler had a characteristic "binder demand" that was a function of the difference in particle size between the titanium pigment and the filler, and the particle-size distribution of the filler. The volume relationship of pigment-to-binder which just satisfied this binder demand was described as the critical PVC, or CPVC, PVC being the commonly used abbreviation for pigment volume concentration. It was possible to identify the CPVC for any combination of pigments by determining the minimum amount of linseed oil that would form a dry but coherent mass when added to a given weight of the pigment and worked vigorously with a spatula—the so-called spatula rub-out oil absorption test (ASTM Method D281).

$$\text{CPVC} = \frac{\text{vol. of pigment}}{\text{vol. of pigment} + \text{vol. of oil}}$$

Furthermore, by varying the percentages of fine and coarse components, it was possible to show the variation of packing with composition on a CPVC curve and to identify the condition of maximum packing (5). This is illustrated in Figure 2 for rutile titanium dioxide in combinations with three calcium carbonate fillers of varying particle size. The coarser the filler, the more pronounced the peak in the curve representing

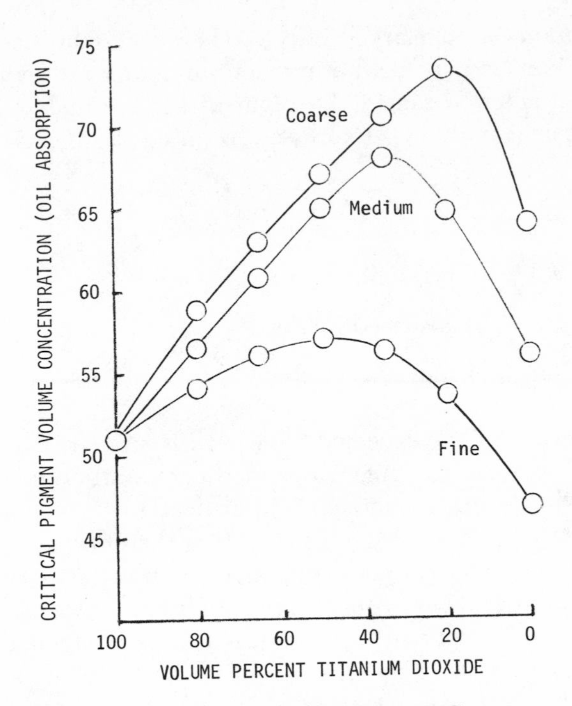

Figure 2. Calcium carbonate CPVC curves

a condition of maximum packing. For titanium dioxide and most large particle size fillers, this peak typically occurs at a concentration (v/v) of about 20% titanium dioxide.

The evidence of such oil absorption end points can also be interpreted in terms of the maximum amount of filler that can be incorporated into a fixed amount of binder.

$$\text{gal filler/gal resin} = \frac{\text{CPVC}}{(1\text{-CPVC})}$$

For the coarser calcium carbonate of Figure 1, for example, the total loading for the pure filler can be increased from 1.82 gal per gal of resin to 3.02 gal by using a blend with 20% TiO_2. This amounts to an increase of almost 66% in maximum loading. The discovery that oil absorption effects were not simply additive provided an explanation for many phenomena that had puzzled paint formulators.

Particle Size Distribution

Average particle size and CPVC were not sufficient information, however, to explain the shape of many of these CPVC curves. Particle-size distribution had a marked effect. The CPVC had been defined by Asbeck and Van Loo (*6*) as that pigment-to-binder ratio which provided just enough binder to satisfy the pigment surface and to fill the voids between pigment particles. Expressed as an equation, this became (*7*):

$$\mathrm{CPVC} = \frac{P}{P+a+b}$$

where P = vol % of pigment

a = vol % of adsorbed binder

b = vol % of binder in voids

The term a could be expected to vary with surface area and therefore with average particle size, but b depended upon particle-size distribution. Together $a + b$ were numerically equivalent to the volume of binder required to wet P volumes of pigment. It is therefore quite possible for two very different fillers to possess similar oil absorptions (and CPVC's). A large-particle-size filler with a very uniform particle-size distribution will have an oil absorption made up of a small a component but a large b component (because of the absence of packing), while a fine-particle-size filler with a wide particle-size distribution would have an oil absorption composed of a large a component and a small b component. Yet the sum of $a + b$ could theoretically be the same for both. The CPVC curve would disclose the difference, however, because the large voids of the coarser filler would permit a much higher percentage of titanium dioxide to be packed into them while the voids of the finer filler would be essentially filled to begin with.

Table I. Effect of Particle-Size Distribution

Type of Filler	*Oil Absorption*[a]	
	No TiO_2	*20%* TiO_2
Unclassified $CaCO_3$	17.3	14.0
Classified $CaCO_3$	18.8	10.6

[a] Lbs of oil per 100 lbs of pigment (ASTM Method D281).

The difference in potential packing is illustrated in Table I for two calcium carbonates—one a normal dry ground material, the other a classified fraction of the first with the fines removed. Both could be called coarse—in the 5–10 micron average particle-size range—but the classified material had a narrower particle-size distribution.

The unclassified filler had a lower oil absorption than the classified material because of the packing of its wider range of particle sizes. The addition of 20% titanium dioxide produced a lower oil absorption with the classified material, however, because of the availability of larger voids not present in the unclassified material because of packing. The titanium dioxide producing this effect had an oil absorption of 21.2, yet it was able to increase packing because of its smaller particle size when added to the coarser extenders.

Since coatings are sold on a volume basis, the volume relationship CPVC is of more interest than the oil absorption weight relationship. The CPVC in coatings is closely associated with a transition point for many film properties, among them washability and stain removal, enamel hold-out, color uniformity over surfaces of varying porosity, vapor permeability, and blistering. Consequently, control of filler particle size has become a significant factor in paint formulation.

Maximum Packing

The relationship between pigmentation CPVC and PVC which has been described as the porosity index, PI, has been used to predict the dry hiding power of porous coatings, as well as film porosity (8, 9):

$$PI = 1 - \frac{\text{CPVC (1-PVC)}}{\text{PVC (1-CPVC)}}$$

In liquid coatings with no pigment/binder interactions, viscosity has been shown to be related to the packing characteristics of the pigmentation, as well as to its volume concentration, by the equation:

$$\log \eta_R = \frac{P}{\text{CPVC-}P}$$

where η_R = relative kinematic viscosity of pigment paste to clear vehicle

P = percent pigment by volume

Because of this relationship of viscosity to the CPVC—and therefore to pigment packing—it is possible to increase the maximum amount of pigment that can be incorporated in a given vehicle by deliberately selecting a combination of large and small particle size pigments (and/or fillers) to produce maximum packing. By a similar technique, viscosity can be reduced at a fixed pigment loading.

In plastics systems where the volume of titanium dioxide used is very small compared with coatings, the use of a blend of large and small particle size filler may be desirable. While low oil absorption may have been the basis for originally selecting a particular filler, it is quite possible, indeed probable, that its low oil absorption is more the result of very large particle size (and therefore low surface area) than of a particle size distribution specifically designed to reduce void volume.

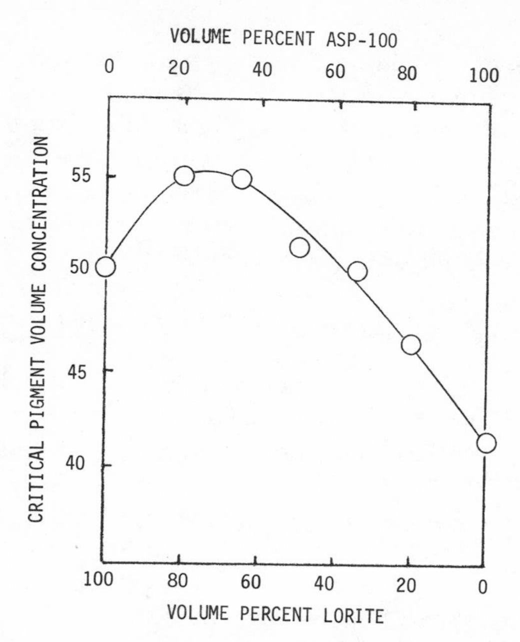

Figure 3. CPVC curve for filler blend

By dry blending fillers in various proportions by volume and determining the spatula rub-out oil absorption for the blends, it is possible to identify the combination of maximum packing, as illustrated in Figure 3. For the fillers involved in this experiment, the CPVC of maximum packing (55.4) represents an increase of almost 23% in the amount of filler that may be used per 100 lbs of resin compared with the lower oil absorption filler (Lorite, a product of NL Industries, Inc.) alone. As in coatings, the particle size of the largest filler used may be limited by film thickness, surface texture, mill abrasion, or other considerations. This imposes a lower limit on the oil absorption of the filler since, as a general rule, oil absorption tends to increase as particle size is reduced. It will usually be possible, however, to find an even finer filler that will reduce the oil absorption when added in small amounts. An excellent source of data on such fillers is the "Raw Materials Index, Pigment Section," published by the National Paint and Coatings Association, 1500 Rhode Island Ave., N.W., Washington, D. C.

Literature Cited

1. ASTM Standard D883, Committee D-20, ASTM, Philadelphia, Pa.
2. Gardner, H. A., "Paint Technology and Tests," p. 105-200, McGraw-Hill, New York, 1911.
3. Calbeck, J. H., *Ind. Eng. Chem.* (1926) **18,** 1220.
4. Stieg, F. B., *Off. Digest* (1957) **29,** 439.
5. Stieg, F. B., *Off. Digest* (1956) **28,** 695.
6. Asbeck, W. K., Van Loo, M., *Ind. Eng. Chem.* (1949) **41,** 1470.
7. Stieg, F. B., *J. Paint Technol.* (1967) **39,** 703.
8. Stieg, F. B., Ensminger, R. I., *Off. Digest* (1961) **33,** 792.
9. Stieg, F. B., *J.O.C.C.A.* (1970) **53,** 469.

RECEIVED October 11, 1973.

3

Asbestos as a Reinforcement and Filler in Plastics

JOHN W. AXELSON

Johns-Manville Research Center, P. O. Box 5108, Denver, Colo. 80217

Four varieties of asbestos are used in the plastics industry: chrysotile, crocidolite, amosite, and anthophyllite. Some are used as raw fiber, felts, cloth yarns, and paper. Asbestos wets out in resin systems but is difficult to disperse. Because of its high chemical reactivity, under certain conditions it tends to cause some high temperature instability with certain resins—e.g., *PVC and polypropylene. Generally, asbestos will improve flexural modulus of plastics 100% or more, flexural strength up to 50%, and heat distortion temperature by 30°F or more. Thermal coefficient of expansion is also reduced. Impact strength can be reduced as much as 50% by adding asbestos. However, this apparently can be overcome by using lubricants or bond breakers on the asbestos to minimize the resin–asbestos bond; flexural properties and heat distortion temperature are not affected.*

Asbestos has been recognized as an individual material from the time of Marco Polo—*i.e.*, for more than 700 years. However, its widespread use did not occur until the twentieth century. Today's estimated worldwide asbestos use is almost five million tons. Canada produces almost 40% of this amount and Russia almost one-half.

What is asbestos? The name is a generic term given to a group of fibrous minerals. The largest portion contains only one mineral which is called chrysotile from the serpentine group and is the white asbestos comprising about 95% of the world production. The second portion of commercial asbestos is in the amphibole group and has three varieties of importance. The first is crocidolite or blue asbestos which is used as a reinforcing fiber in some plastics because of its acid resistance. It is available as fiber, paper, yarn, felt, and cloth. Chrysotile is also available in these forms. Amosite is the second amphibole but is not used in

Table I. Chemical Formulas of Asbestos

Chrysotile	$Mg_6[(OH)_4Si_2O_5]_2$
Crocidolite	$Na_2MgFe_5[(OH)Si_4O_{11}]_2$
Amosite	$MgFe_6[(OH)Si_4O_{11}]_2$
Anthophyllite	$(Mg, Fe)_7[(OH)Si_4O_{11}]_2$
Tremolite	$Ca_2Mg_5[(OH)Si_4O_{11}]_2$
Actinolite	$Ca_2(Mg, Fe)_5[(OH)Si_4O_{11}]_2$

Figure 1. Plain view of chrysotile

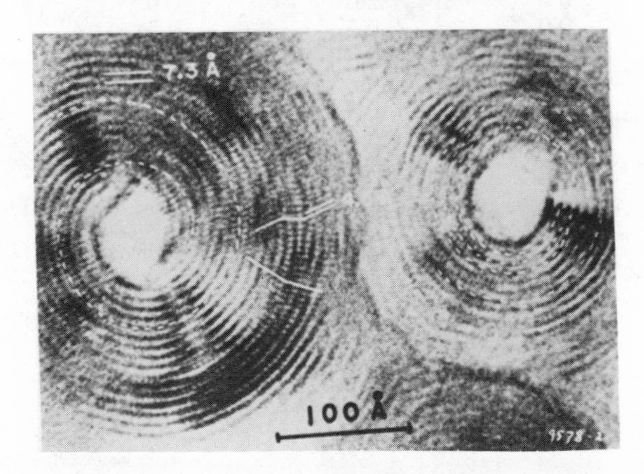

Figure 2. Cross-sectional view of chrysotile

plastics as far as is known. The third amphibole—anthophyllite—available only as a short fiber, is used extensively in filled polypropylene.

Chemical formulas for the two families of asbestos are given in Table I. Figure 1 is a plain view of chrysotile asbestos, and Figure 2 is a cross-sectional view. Both show the hollow, tubular nature of chrysotile.

Asbestos is a naturally occurring mineral largely mined by open pit but with some by underground techniques. The rock, containing veins of fiber, is blasted and crushed to about a 1-inch maximum size. The material then goes through vertical hammer mills called fiberizers to release the fiber. Gyratory screens separate the fiber from the rock. The undersized rock goes to tailings; the oversized rock goes to fiberizers where the action is repeated; the fiber is air aspirated off the end of the screens to cyclone collectors. The fiber may be processed further to remove undesirable fractions or to change its character, but eventually it is graded and bagged for shipment.

Table II. Grading of Asbestos

Grade	*Length or Distribution by Screen Size*
No. 1 Crude	3/4 inch or longer
No. 2 Crude	3/8 to 3/4 inch

	Quebec Screen Analysis			
	1/2 inch	*4 Mesh*	*10 Mesh*	*Pan*
3K	7.0	7.0	1.5	0.5
3T	2.0	8.0	4.0	2.0
4D	0.0	7.0	6.0	3.0
4T	0.0	2.0	10.0	4.0
5D	0.0	0.5	10.5	5.0
5R	0.0	0.0	10.0	6.0
6D	0.0	0.0	7.0	9.0
7D	0.0	0.0	5.0	11.0
7R	0.0	0.0	0.0	16.0

Table III. Approximate Comparative Lengths and Costs of Chrysotile

Grade	*Approximate Comparative Length, inch*	*Approximate Delivered Price, ¢/lb*
2	5/8	45
3	1/2	27
4	3/16	13
5	1/8	10
6	1/16	7
7	1/32	4

Asbestos grading is generally done by a dry screen analysis. The most popular is the Quebec screen test which designates the maximum ounces on the top screen and the minimum ounces in the pan for a 1-lb sample. The test procedure is very specific, and typical sample grade designations are given in Table II. The shorter the length of the fiber, the higher the first digit in the classification; the lower in the alphabet

Table IV. Properties of Asbestos

Property	*Chrysotile*	*Crocidolite*	*Amosite*	*Anthophyllite*
Color	white to gray	blue	brown	gray to brown
Tensile strength, psi	281,000–436,000	469,000–605,000	148,000–203,000	350,000
Modulus of elasticity, psi	23.2×10^6	27.1×10^6	23.6×10^6	22.5×10^6
Hardness, mohs	2.5–4	4	5.5–6.0	5.5–6.0
Flexibility	good	fair	poor	poor
Specific heat, Btu/lb/°F	0.266	0.201	0.193	0.210
Specific gravity	2.4–2.6	3.2–3.3	3.1–3.25	2.9–3.2
pH	10.3	9.1	9.1	9.4
Refractive index	1.50–1.55	1.70	1.64	1.61
Fibril diameter, A	160–300	600–900	600–900	600–900
Surface area, BET, m^2/gram	1.7–60	9–10.5	8–9	6–7
Coeff. of cubical exp, °F	5×10^{-5}			
Charge in water	positive	negative	negative	negative
Isoelectric point	11.3–11.8			
Resistance to acids	poor	good	good	good
Resistance to bases	good	good	fair	good

Table V. Mechanical Properties of Asbestos-Reinforced Resins

Material	*Tensile Strength, psi*	*Flexural Strength, psi*	*Tensile Modulus, psi* ($\times 10^6$)
Phenolic crocidolite felt	30,000	58,000	3.3
Polyester crocidolite felt	35,000	45,000	3.0
Melamine formaldehyde chrysotile paper	9,000	18,000	1.7
Phenolic chrysotile molding compound	13,000	30,000	4.0

in a grade series, the shorter the fiber. Additional suffixes by each producer indicate the degree of openness of the fiber. Table III lists approximate average lengths for the various grades of asbestos fiber and also an approximate delivered price.

One of the attributes of asbestos fiber is its relatively low price *vs.* its performance. Naturally, the price is reflected by the length of the fiber. Later, we show that the shorter, lower priced fibers are generally those that are used in plastics with good effectiveness.

Asbestos exhibits unique properties (Table IV). One property not listed in Table IV is its ability to be wet out by all resin and latex systems. Occasionally it is difficult to disperse asbestos properly because the individual fibrils are very small and tend to agglomerate, but there is no evidence that asbestos is not easily wetted out by all systems. One of the unique properties of chrysotile is the positive charge it develops in the presence of water; this may have some bearing on its excellent wettability.

Asbestos fiber can be considered as both a reinforcing fiber and a reinforcing filler. In the first case, special long grades of asbestos or special forms such as paper, cloth, or felt are used to obtain high performance products. Some of the properties which can be obtained are shown in Table V. As a reinforcing filler, asbestos imparts many properties which are not obtained with granular fillers. Various advantages and disadvantages of asbestos as a reinforcing fiber or filler are shown in Tables VI and VII.

Table VI. Asbestos as a Reinforcing Fiber

Advantages	*Disadvantages*
Low cost	Fair impact strength
Easy flash removal	Dark color
Flow control in mold	Possible mixing difficulties
Mininum fiber degradation	
Good flexural strength	
Low water absorption	
Heat resistance	
Good electrical properties	
Chemical resistance	

Table VII. Asbestos as a Reinforcing Filler

Advantages	*Disadvantages*
High modulus	Increase in resin viscosity
Tensile improvement	Fair electrical resistance
Flow control in mold	Some abrasiveness
Good surface finish	Possible polymer degradation
Heat stability	High density
Dimensional stability	
Low creep	
Chemical resistance	
Higher hardness	
Arc resistance	
Low water absorption	
Low thermal coefficient	

One controllable aspect of asbestos that should be fully recognized is its effect on the heat stability of certain polymers. Because of its high surface area and chemical reactivity, chrysotile particularly tends to decrease the heat stability of certain polymers such as poly(vinyl chloride) and polypropylene. However, with proper stabilizers and antioxidants.

both polymers can be stabilized for processing temperatures and long term use at elevated temperatures.

The next tables and figures show what kind of physical properties can be obtained with various asbestos fibers. Standard techniques were used to prepare all specimens, and testing was done in accordance with ASTM procedures. Figures 3, 4, and 5 depict the impact and flexural strengths which can be obtained in a polyester mix with asbestos fibers of various lengths (1/8 to 1/2 inch) *vs.* a commercial fiber with a nominal length (3/16 inch). The asbestos fibers for these tests were prepared

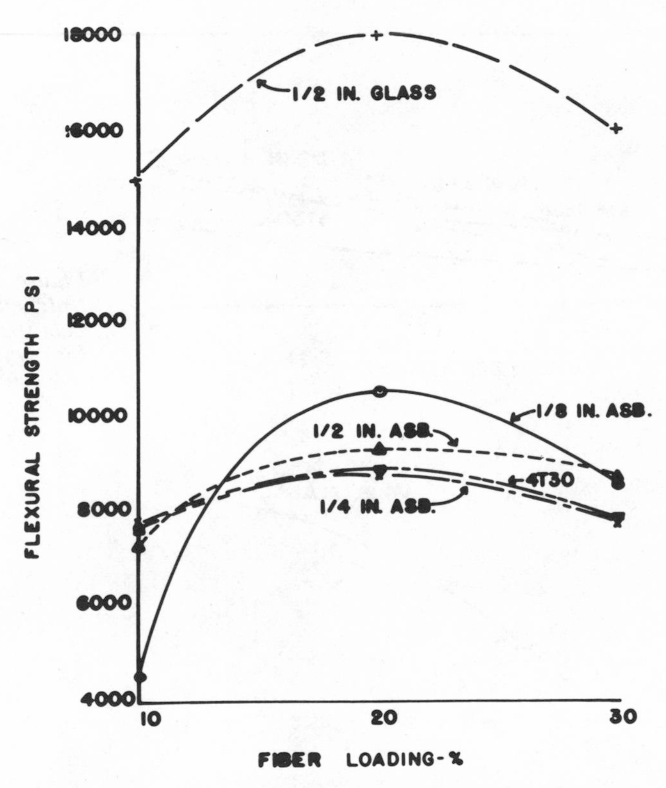

Figure 3. Flexural strength for chrysotile-polyester

from crude fiber. These were carefully cut by hand to the desired length and then opened gently to minimize fiber shortening by passing through a small impact mill. The fibers were incorporated into the liquid polyester in a small sigma blade type mixer. The final mix was compression molded at 300°F for 5 min at 500 psi, and test specimens were cut from the molded piece. As can be seen, fiber length does not seem to have any significant effect on the physical properties obtained.

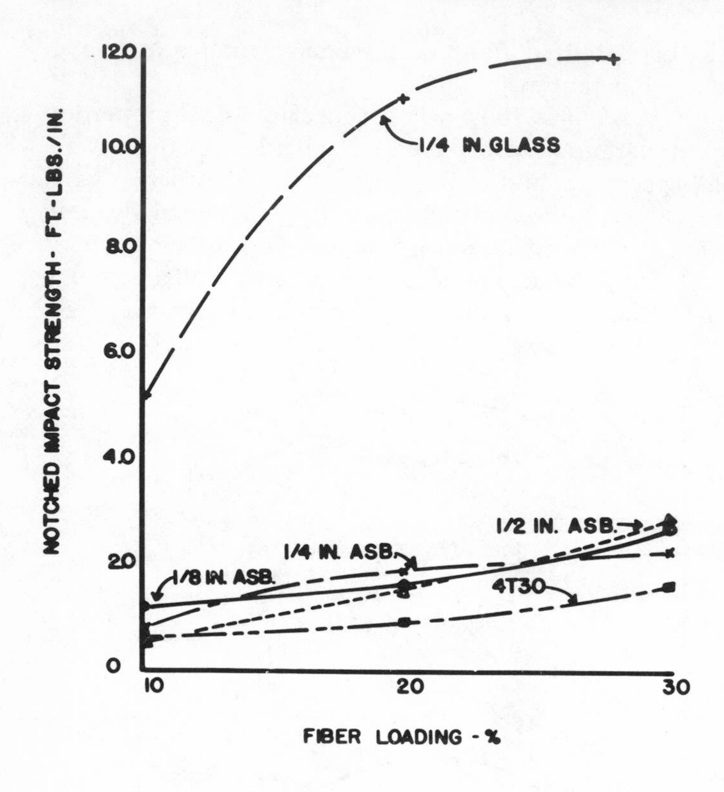

Figure 4. Notched impact strength for chrysotile-polyester

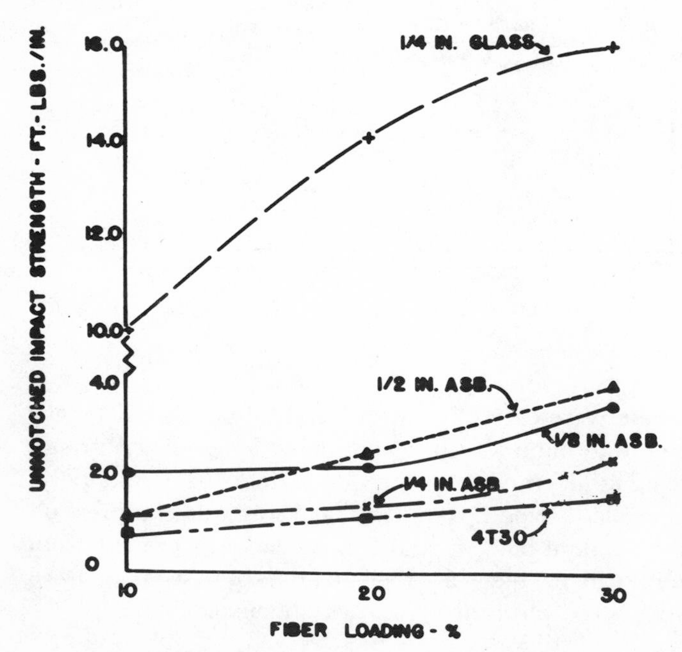

Figure 5. Unnotched impact strength for chrysotile-polyester

Table VIII lists the general treatments which were given to asbestos fiber before incorporation in a polyester resin. The treatments were generally done in a solvent slurry with subsequent drying and opening of the asbestos. Sample preparation and testing were as previously described. Physical properties were not improved with any of these treatments. This adds to the evidence for good fiber wetting and the lack of need for fiber treatment.

Tables IX, X, XI, and XII and Figures 6, 7, and 8 depict some properties obtained when asbestos fiber is incorporated in four typical resins.

Table VIII. Fiber Treatments

1. Crosslink asbestos to resin
2. Flexibilize the fiber to resin bond
3. Coat asbestos to improve compatability
4. Encapsulate asbestos to give integral bundles
5. Let free water react on asbestos surface

Table IX. Asbestos in Polyethylene

Fiber	*% Added*	*Flexural Strength, psi*	*Flexural Modulus, psi* $\times 10^5$	*Tensile Strength, psi* $\times 10^5$	*Impact Strength (Notched Izod) ft-lbs/inch*	*Heat Distortion, °F*
None	—	4170	1.54	3260	0.9	120
7M02	20	4560	2.82	2440	0.4	131
	30	5840	4.21	3380	0.6	157
	40	5440	6.44	3900	0.9	192
Glass, 1/4 inch	20	4800	3.50	3220	1.6	122

Table X. Asbestos in ABS

Asbestos Added	*Flexural Strength, psi*	*Flexural Modulus, psi* $\times 10^6$	*Tensile Strength, psi*	*Impact Strength, (Notched Izod), ft-lbs/inch*	*Distortion, °F*
—	10,650	0.54	5855	1.42	190
40% long fiber (Plastibest #20)	8,765	1.11	4715	0.52	202
40% short fiber (7D04)	8,290	1.39	6340	0.58	208
20% 1/4-inch glass	10,350	0.58	5595	0.70	207

Table XI. Asbestos in Phenolics

Material	*Flexural Strength, psi*	*Flexural Modulus, psi × 10⁶*	*Impact Strength (Notched Izod), ft-lbs/inch*
A. Two Stage Phenolic Powder plus Asbestos			
None	8,900	0.78	0.28
40% 7RF02 fiber	13,500	1.44	0.20
60% 7RF02 fiber	12,900	2.03	0.39
B. Filled Commercial Molding Compound			
General purpose wood flour	13,700	1.03	0.32
Heat resistant J-M 7T15 asbestos	9,800	1.31	0.30

Table XII. Asbestos in Polystyrene

Asbestos Added	*Flexural Strength, psi*	*Flexural Modulus, psi × 10⁵*	*Tensile Strength, psi*	*Impact Strength (Notched Izod), ft-lbs/inch*	*Heat Dostortion, °F*
None	5195	2.91	2585	1.42	179
40% long fiber (plastibest #20)	5180	6.24	3570	1.09	201
40% short fiber (7D04)	7410	5.97	2500	0.56	194
20% 1/4-inch glass	5895	4.86	2680	0.65	194

The polyethylene samples for Table IX were prepared by preblending the resin and fiber in a tumbler and then fluxing on a roll mill at 250°F for 10 min. Pieces were cut 4 × 4 inces from the sheet off the mill and compression molded to samples 4½ inches × 4 × 1/8-inch. Hot pressing was done at 350°F for 5 min with a 1-ton load; then the mold was transferred to a water-cooled press where a 20-ton load was applied over 60 sec and held for several minutes. Spacer bars were used to control thickness. Cutting and testing of samples were according to ASTM procedures.

The preparation of ABS samples for the data in Table X started with fusing of the resin on a roll mill at 325°F and slow addition of the fiber. Sheets 8 × 8 inches were cut and placed in a mold to give a sample 9 × 9 × 1/4 inch. They were pressed at 400°F with 100 tons with a warm-up time of 5 min, 10 min for pressing, and 15 min in the cooling press.

Phenolic specimens in Table XI were prepared by dry blending a two-stage resin and the fiber before plasticating on a roll mill at 290°F for 1½ min. The cooled sheet was passed through a hammer mill, and the resultant powder was compression molded for 30 min at 290°F and 3000 psi to give specimens 9 × 9 × 1/8 inch. Polystyrene was handled in the same way as ABS except the roll mill temperature was only 275°F.

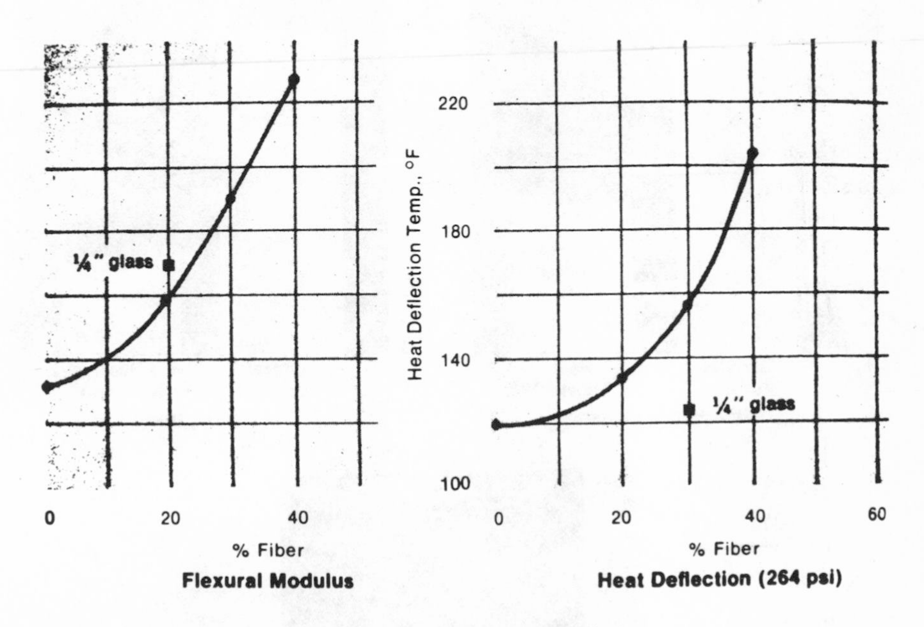

Figure 6. Asbestos in polyethylene

The increases in modulus, flexural strength, and heat deflection temperature are the most notable. Improvement in these properties is the primary reason for using asbestos. Conversely, the major deficiency is the loss in impact strength. It is hoped that this can be overcome by the technique described by Dupont by reducing the bond between the fiber and the matrix (*1*). Some recent work not yet ready for publication shows that this technique does work for asbestos in rigid PVC. The asbestos is pretreated with a lubricant and preblended with PVC before extrusion. Final properties of the extruded article, including impact strength, have equaled those of an unfilled part, and the modulus has shown considerable improvement. This work looks very interesting and should be applicable to other systems.

Recent work with properly stabilized polypropylene has shown the effectiveness of chrysotile *vs.* various other types of fillers. The work reported in Table XIII with polypropylene used the mixing chamber of a Brabender Plasticorder to blend the fillers and resin. This was roughly

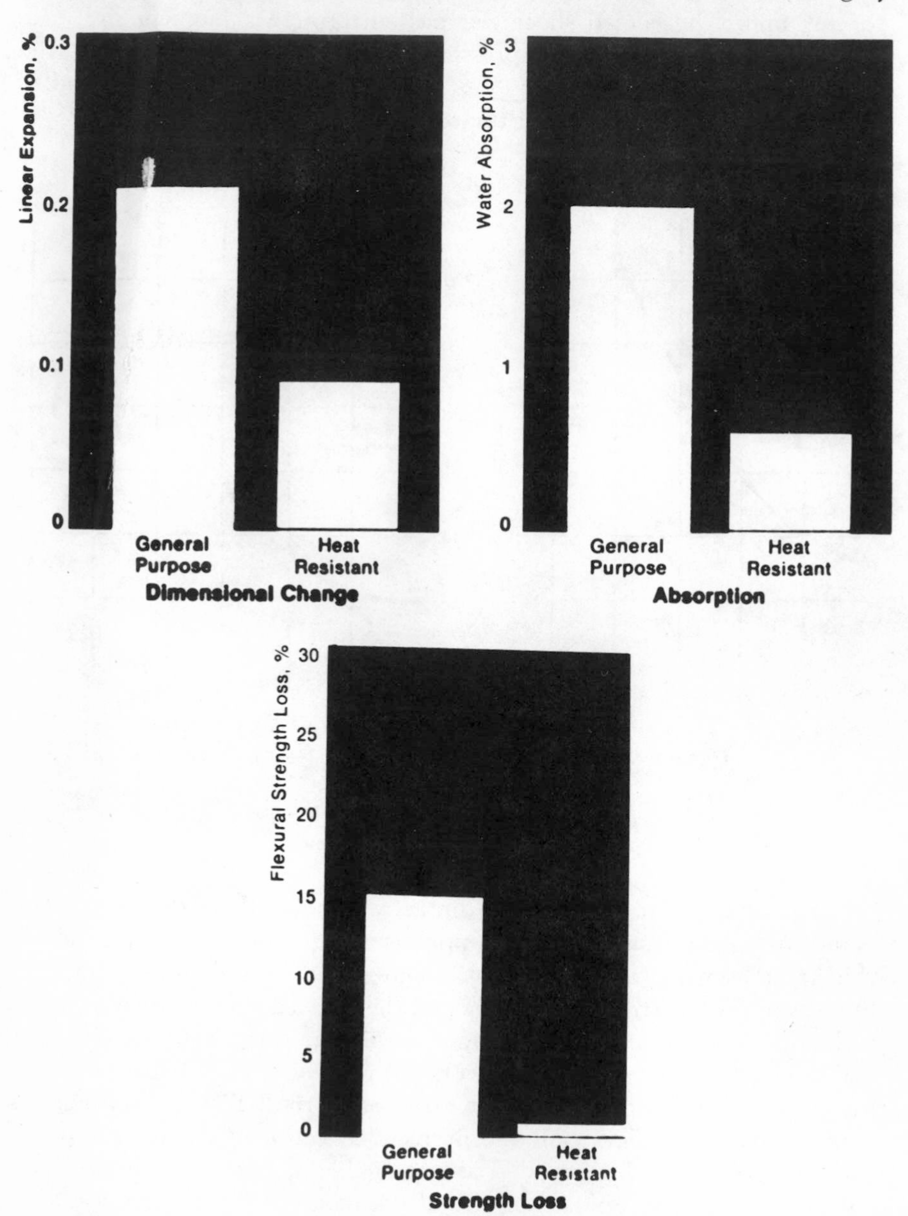

Figure 7. Asbestos in phenolics; two-weeks water saturation at 75°F

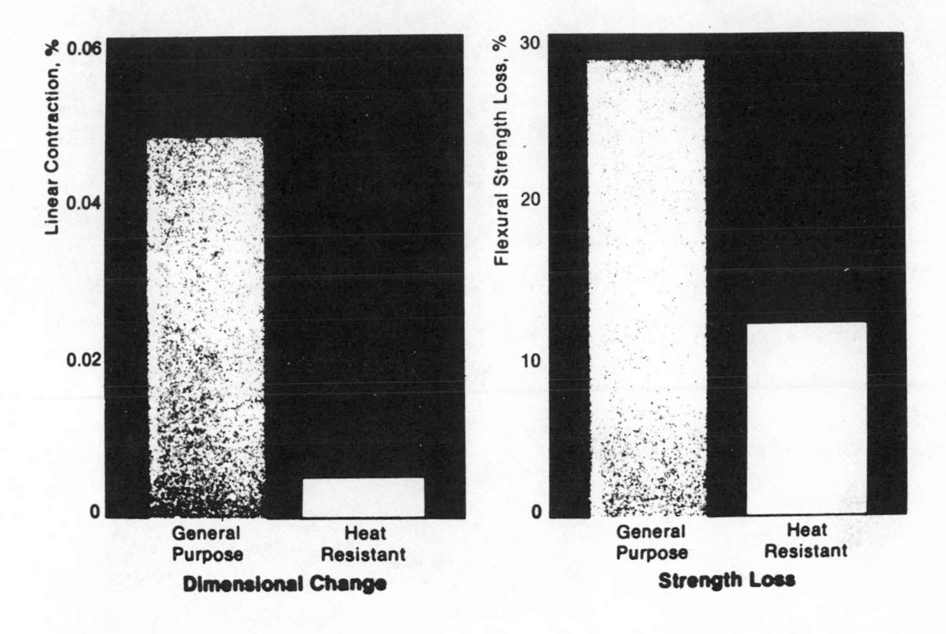

Figure 8. Asbestos in phenolics; two-weeks oven aging at 375°F

Table XIII. Physical Properties of Filled Polypropylene

Filler	*Flexural, psi* *Strength*	*Flexural, psi* *Modulus*	*Heat Deflection Temp, °F*	*Impact Strength (Notched), ft-lbs/inch*	*Heat Stability, hrs at 150°C*
Unfilled	5200	1.9×10^5	136	0.5	2000+
Talc	7200	5.9×10^5	204	0.5	2000+
Wollastonite	5700	3.7×10^5	159	0.5	2000+
Anthoplyllite	6900	5.1×10^5	183	0.5	1000
Chrysotile	9000	6.1×10^5	239	0.5	1500[a]

[a] Contained special antioxidant.

Table XIV. Radiant Panel Flammability Test for High Density Polyethylene–40% Filler

Filler	*Heat Evolution Factor, Q*	*Flame Spread Factor,* F_s	*Flame Spread Index,* $I_s = F_sQ$
$CaCO_3$	10.8[a]	19.6	213
Chrysotile	14.4	4.3	62

[a] Low because of drippage through screen support.

compacted into a disc about 1/4-inch thick and then compression molded using the hot-cold press technique described for polyethylene. However, the hot press temperature was 400°F. One significant feature is the fire resistance which is imparted to an asbestos-filled resin as shown in Table XIV.

Literature Cited

1. Speri, W. M., Jenkins, O. F., "Effect of Fiber-Matrix Adhesion on the Properties of Short Fiber Reinforced ABS," *Ann. Tech. Conf., 28th,* Society of Plastics Industry, Washington, D. C., 1973.

RECEIVED October 11, 1973.

4

Asbestos-Reinforced Rigid Poly(vinyl chloride)

A. CRUGNOLA and A. M. LITMAN[1]

Lowell Technological Institute, Lowell, Mass. 01854

Two rigid poly(vinyl chloride) polymers (a low and a high molecular weight sample) were reinforced at several concentrations with four chrysotile asbestos fibers of varying length and openness. Mechanical, thermal, and other physical tests were done on the resultant composites: load/elongation curves, impact, hardness, heat deflection temperature, and flammability. Complete analyses of the before and after processed polymers' molecular weight and molecular weight distribution were also done. X-ray radiographs of the composites established fiber distribution in the molded specimens; infrared analyses on the polymers showed possible changes in chemical structure. A new mixing technique incorporated the asbestos in the plastic. Resultant composite properties generally depended on four factors. In decreasing importance they are: (1) matrix polymer molecular weight, (2) fiber concentration, (3) fiber "openness," (4) fiber length.

Little research has been done on the reinforcing effects of asbestos on rigid poly(vinyl chloride) (PVC) plastics (*1*). This combination is interesting because of the inherent properties of PVC and the unique characteristics of asbestos fibers. Rigid (unplasticized) PVC is a low cost, tough thermoplastic noted for excellent chemical and flame resistance and extensive processing versatility—properties which lead to countless applications in both the building and the chemical areas. Two such polymers were used in this work: a low molecular weight product of $\overline{M}_w$ = 49,900 and a high molecular weight product of $\overline{M}_w$ = 116,380.

Asbestos is a broad term applied to a number of natural silicated minerals that are incombustible and can be separated either mechanically or chemically into fibers of varying lengths and thicknesses. Low cost,

[1] Present address: Army Materials and Mechanics Research Center, Watertown, Mass. 02172.

excellent heat resistance, and exceptional mechanical behavior account for the growing use of asbestos in the reinforced plastics industry. There are two distinct mineralogical classifications of asbestos: those from serpentine rock formations and those from amphibole rock formations.

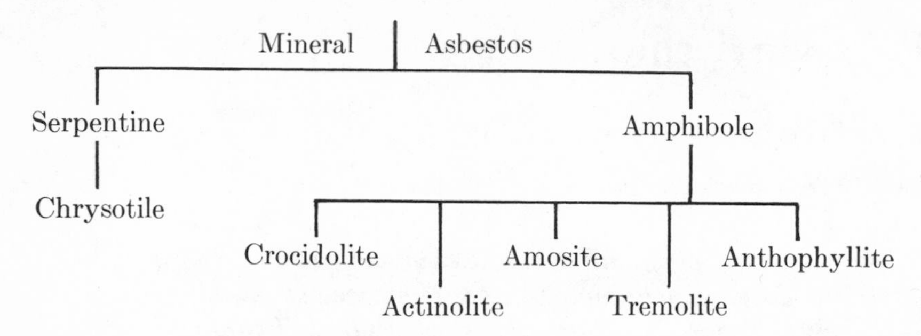

Of the six varieties cataloged only four have commercial use as reinforcements: chrysotile, amosite, anthophylite, and crocidolite. The chrysotile asbestos accounts for about 95% of world production and is the most widely used in plastics. The advantages of this fiber as a reinforcement are: (1) higher strength and modulus *vs.* glass fibers, (2) a positive surface charge which promotes fiber/matrix bonding, (3) availability in a wide range of lengths and diameters, (4) a soft, flexible quality which results in minimum wear on processing equipment. The four chrysotile fibers used in this study are described in Table I.

Table I. Asbestos Fibers Used[a]

Fiber	*Size, inch*	*Description*
7RS7	1/32	short standard fiber used in thermoplastic systems
Paperbestos No. 5	3/64	shorter version of Paperbestos No. 1
Paperbestos No. 1	1/4	clean, well opened fiber of good length
Plastibest No. 20	1/8-3/16	clean, medium length fiber not opened very much

[a] Fibers supplied by Johns Manville Co. and classification/descriptions are theirs.

Experimental

Specimen Fabrication. A measured amount of PVC was placed in a Henchel mixer and mixed at maximum speed (3600 rpm) until the batch temperature reached 200°F. At this temperature the stabilizer was added (2 pph by weight of Advances' T-360, a sulfur-containing organotin compound), and mixing continued until the temperature reached 250°F; at this time the material was discharged into a ribbon blender for cooling (approximately 5 min).

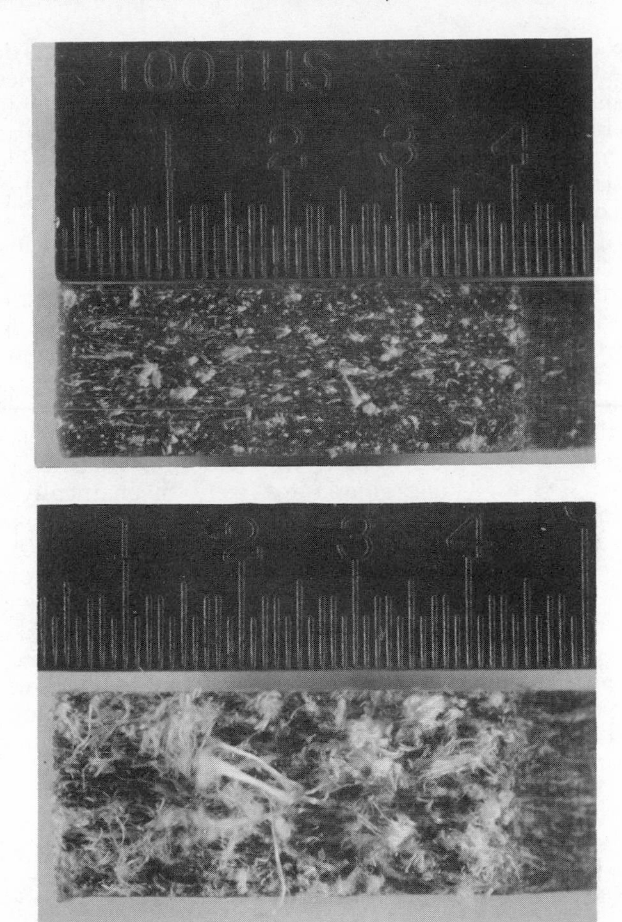

Figure 1. Impact fracture surfaces (notched Izod). Top: Paperbestos No. 5/PVC; bottom: Plastibest No. 20/PVC.

The asbestos was dried in a non-circulating air oven at 250°F for 24 hrs to remove moisture that otherwise might prevent the fiber bundles from separating in the premixing operation.

The breakup of the asbestos clumps and the dry-blending with the stabilized polymer were done in a specially designed mixing device consisting of a glass jar, a ball mill, and a nitrogen source. The mixing jar was equipped with a perforated shaft running down its length which permitted the nitrogen introduced under a nominal pressure to be uniformly directed throughout the jar. A Teflon bearing seated at the center of the jar cover permitted the bottle to turn on the ball mill independent of the shaft.

The PVC/asbestos mixtures obtained by dry-blending were milled on a Farrell two roll mill at 360° ± 5°F. Small quantities (approximately 200 grams at the time) were fluxed on the 15-inch long rolls to ensure thorough fusion and mixing in a short time. The mixing/fluxing times depended somewhat on fiber type and content and varied between 1½ and 2 min. Despite the short times used, there was no evidence of incomplete fusion either during fluxing or on viewing the impact fracture surfaces. This is not to say that every fiber was wet with polymer which wetting instead depended on the degree of fiber openness. This is vividly demonstrated in the photographs of Figure 1.

Finally the composite panels were compression molded in a Wabash press at 2000 psi. Several sheets from the milling operation were cross-plied, placed in an open frame mold and brought to temperature (355° ± 5°F for the high molecular weight matrix and 330° ± 5°F for the low molecular weight matrix) before applying pressure. The pressure was maintained for 1 min hot, then the frame was transferred to a cold press and the material cooled under the same pressure (2000 psi). Composites of each of the four asbestos fibers were fabricated at 15, 30, and 45% fiber content in each of the two poly(vinyl chloride) plastics.

Specimen Testing. The composites were mechanically tested for tensile modulus, breaking elongation, breaking strength, impact, hardness, and heat deflection. ASTM flammability evaluations were also performed. The effectiveness of the specimen preparation technique with regard to the distribution of the fibers within the composite was established by x-ray radiographs of selected samples. The polymer/matrix materials were studied by (1) gel permeation chromatography to follow changes in molecular weight and molecular weight distribution caused by the processing of the fiber/plastic mixtures into the composites and (2) infrared analysis in an attempt to pick up possible changes in the chemical structure of the polymers.

Table II. Tensile Modulus

	7RS7		*No. 5*	
Molecular Weight	*High*	*Low*	*High*	*Low*
% Asbestos Fiber				
0	—	4.24[b]	—	4.24
15	5.58	5.27	5.47	6.46
30	6.75	7.82	6.65	6.70
45	8.25	8.03	8.07	7.28

[a] Strain rate 4.5%/min; variation in values used to obtain average value = ± 4%.
[b] $\times 10^{+5}$ psi.

Results and Discussion

The results of the testing done on the composites and their components are presented in Tables II, III, IV, V, and VI and in Figures 2, 3, 4, and 5. The findings are summarized below.

Tensile Modulus. Introduction of asbestos fiber raised the moduli of the PVC plastics. The increase generally increased with fiber content and was as much as 100% at 45% reinforcement. Usually the higher molecular matrix plastics exhibited higher moduli than the corresponding low molecular weight materials although this effect was not consistent.

No clear effects of either fiber length or fiber openness were apparent in the results.

Tensile Strength. An increase in tensile strength was generally noted upon the initial introduction of asbestos fibers in PVC plastics. This increase peaked, and then the strength decreased at the higher fiber content levels to the point where the reinforced samples exhibited breaking stresses lower than those of the nonreinforced materials. Usually the higher molecular weight matrix composites showed somewhat higher strengths than did the corresponding low molecular weight specimen. The reason for the unexpected lowering at higher fiber contents is not clear unless the higher fiber concentrations brought with them a higher concentration of internal stress concentrating flaws in the compression molded sheets.

Breaking Strain. Although the nonreinforced plastics at this strain rate (4.5% per minute) were prone to necking and drawing, once the fibers were introduced the breaking strains fell to about 2% (for 15% fiber). Further increases in fiber content further decreased the elongation to break to levels of the order of 1%. No clear effect of fiber length or fiber openness was seen. In general, up to 30% fiber content, the higher molecular weight samples exhibited somewhat higher breaking strains.

Impact Strength. The most important factor influencing the impact strength was the extent to which the asbestos fibers were opened. The

of PVC Asbestos Composites[a]

No. 1		*No. 20*	
High	*Low*	*High*	*Low*
—	4.24	—	4.24
6.65	5.94	5.80	5.15
6.86	5.54	6.37	6.20
5.57	6.52	7.53	7.66

closed fiber structures always gave higher impact strengths. Somewhat higher strengths were associated with the higher molecular weight matrix plastic, the greater fiber concentrations, and the longer fiber lengths. Photographs of the fracture surfaces of a closed fiber and an open fiber reinforced PVC appear in Figure 1. These, we believe, reveal the mechanism by which the non-opened fibers enhance the impact strength of the composite. Considerable energy must be involved in pulling apart the asbestos fibers during the impact—energy used in overcoming frictional forces, *i.e.*, forces required to pull individual fibers away and past others. One might visualize these forces in a model where a fasces structure is pulled apart by grasping some of the rods on one end and some on their

Table III. Tensile Breaking Stress

Molecular Weight	*7RS7*		*No. 5*	
	High	*Low*	*High*	*Low*
% Asbestos Fiber				
0	6,256[b]	4,656	6,256	4,656
15	5,756	5,251	6,978	5,572
30	6,347	4,726	7,270	5,889
45	5,787	4,679	5,734	5,435

[a] Strain rate 4.5%/min; variation in values used to obtain average value = ± 4%.
[b] Psi.

Table IV. Tensile Breaking Strain

Molecular Weight,	*7RS7*		*No. 5*	
	High	*Low*	*High*	*Low*
% Asbestos Fiber				
0	187[b]	67	187	67
15	1.2	2.0	2.6	1.7
30	1.4	1.2	1.8	1.2
45	0.9	0.9	0.8	0.9

[a] Strain rate 4.5%/min.
[b] Percent strain.

Table V. Molecular Weight before and after Processing/Molding

Polymer	$\overline{M}_N$	$\overline{M}_W$	$\overline{M}_W/\overline{M}_N$
SCC—686[a]	56,570	116,380	1.98
O—H—O[b]	52,730	129,500	2.46
1—H—30[c]	58,630	125,700	2.14
7—H—30[d]	46,140	125,100	2.71

[a] Polymer before processing (non-reinforced).
[b] Polymer extracted after processing/molding (non-reinforced).

Table VI. Molecular Weight before and after Processing/Molding

Polymer	$\overline{M}_N$	$\overline{M}_W$	$\overline{M}_W/\overline{M}_N$
SCC—600[a]	23,500	49,900	2.11
O—L—O[b]	20,320	48,200	2.38
1—L—30[c]	21,880	50,340	2.30
7—L—30[d]	22,640	48,190	2.13

[a] Polymer before processing (non-reinforced).
[b] Polymer extracted after processing/molding (non-reinforced).

of PVC/Asbestos Composites[a]

No. 1		No. 20	
High	*Low*	*High*	*Low*
6,256	4,656	6,256	4,656
7,343	5,406	6,319	5,508
5,315	4,349	6,423	3,876
3,709	3,748	3,618	4,719

of PVC Asbestos Composites[a]

No. 1		No. 20	
High	*Low*	*High*	*Low*
187	67	187	67
2.4	1.6	2.2	1.7
1.6	1.0	2.1	1.2
0.8	0.7	0.7	0.9

with Chrysotile Asbestos Fibers for High Molecular Weight PVC

$\overline{M}_Z$	$\overline{M}_Z/\overline{M}_W$	$\overline{M}_{Z+1}$	$\overline{M}_{Z+1}/\overline{M}_Z$
186,600	1.61	249,100	1.30
234,200	1.81	350,800	1.30
219,600	1.75	325,000	1.48
238,300	1.91	370,000	1.56

[c] Polymer extracted after processing/molding reinforced with No. 1 asbestos fibers—30% fiber content.
[d] Polymer extracted after processing/molding reinforced with 7RS7 asbestos fibers—30% fiber content

with Chrysotile Asbestos Fibers for Low Molecular Weight PVC

$\overline{M}_Z$	$\overline{M}_Z/\overline{M}_W$	$\overline{M}_{Z+1}$	$\overline{M}_{Z+1}/\overline{M}_Z$
94,120	1.88	159,200	1.69
93,100	1.93	148,800	1.59
94,920	1.89	149,800	1.58
86,620	1.79	133,700	1.54

[c] Polymer extracted after processing/molding reinforced with No. 1 asbestos fibers—30% fiber content
[d] Polymer extracted after processing/molding reinforced with 7RS7 asbestos fibers—30% fiber content

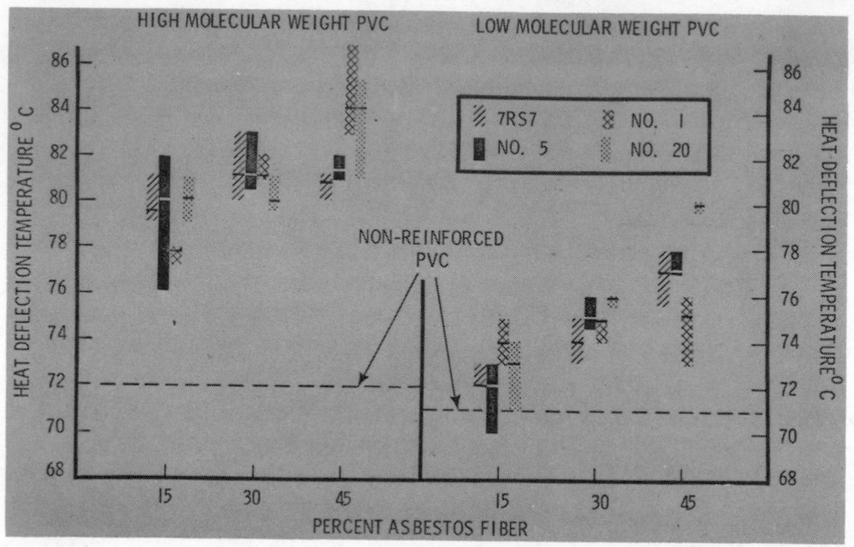

Figure 2. Variation of impact strength with fiber content (reinforced rigid PVC)

other end. The extent of this mechanism apparently depends on the asbestos fiber bundles *not* being thoroughly wet out by the matrix since the better wet open asbestos fibers are fractured at the fracture surface of the polymer.

Heat Deflection Temperature. In all cases the heat deflection temperature was raised by the introduction of the asbestos (6°–12°C for the high molecular weight plastic and 2°–8°C for the low molecular weight material). This temperature increased with increasing fiber con-

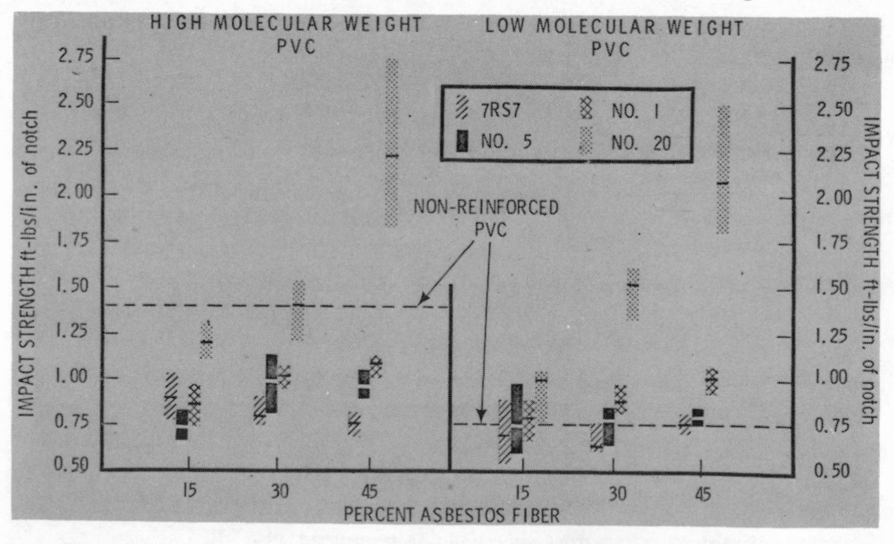

Figure 3. Variation of HDT with fiber content (reinforced rigid PVC)

centration and/or decreasing fiber openness. The molecular weight of the starting plastics proved most important in establishing the amount of improvement in HDT. Whereas the non-reinforced materials differed by only 1°C the reinforced materials differed by as much as 9°C.

Hardness. The asbestos composites consistently showed lower values of hardness than did the non-reinforced plastics. The lowering was greater the greater the concentration of fiber. The somewhat higher hardness of the starting higher molecular weight material appeared to be reflected in the somewhat higher hardness of the higher molecular weight com-

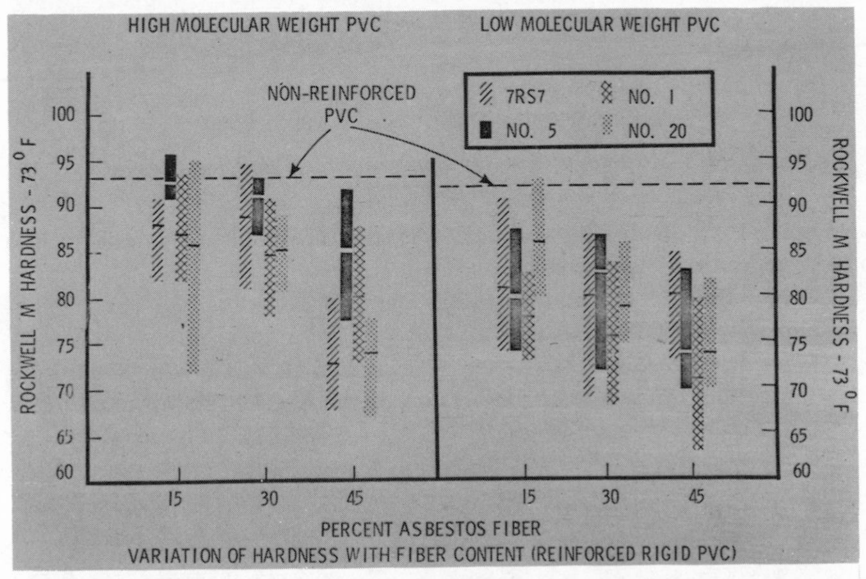

Figure 4. Variation of hardness with fiber content (reinforced rigid PVC)

posites. The lowering of the hardness with the addition of asbestos fibers might be explained on the basis of the low stiffness of these fibers. Despite the fact that the fiber has a very high modulus (*ca.* 20×10^6 psi *vs.* 10×10^6 psi for glass), chrysotile is also an asbestos of finest diameter 7×10^{-7} inch *vs.* 2.6×10^{-4} inch for glass. The hardness test probably subjects the fibers to bending, and as such the stiffness (EI) depends on the fiber diameter as well as the modulus. Since the asbestos is lower in diameter by three orders of magnitude while only higher by a *factor* of 2 in modulus, one would expect to observe lower values of bending stiffness and lower hardness.

The densities of two of the 30% composites were determined, and values of 1.65 and 1.58 grams/cm³ were observed for samples 20-H-30 and 7-H-30, respectively. These compare favorably with a theoretically calculated 1.66 grams/cm³ for this fiber content and suggest that the

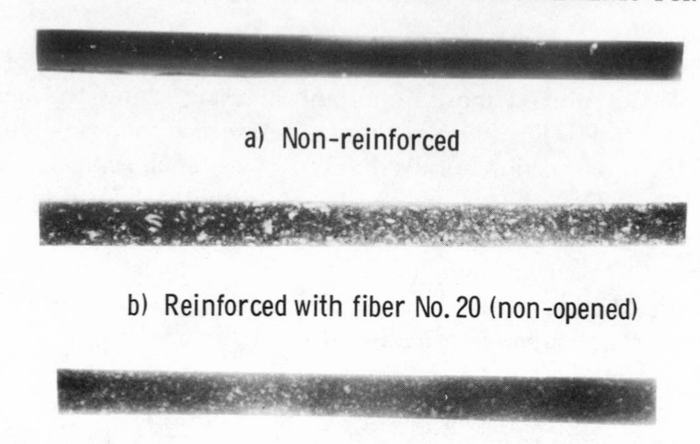

Figure 5. Radiographs of non-reinforced and reinforced PVC

entrapment of air in the composite is not a significant factor in the observed hardness lowering.

Flammability. All the composites were classified as non-burning according to ASTM 635.

Fiber Distribution. As shown by Figure 5 the air mixing technique we used resulted in a uniform dispersion of the fibers in the plastics. This

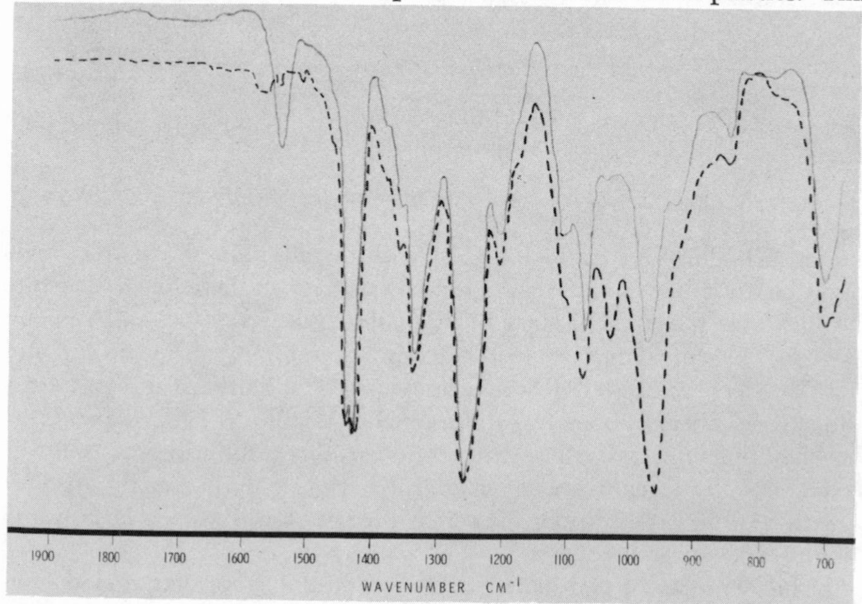

Figure 6. Infrared spectra of polymer before and after processing with asbestos. Solid line: stabilized low molecular weight polymer before processing; dashed line: stabilized low molecular weight polymer after processing with No. 1 asbestos.

was done without affecting the relative degree of fiber openness and/or damaging the fiber length.

Effect of Processing on the Asbestos Fibers. The No. 1 fiber composites were examined for fiber degradation, and fibers did not appear to be significantly shortened by processing. The microscopic view revealed that the asbestos bundles had been sheared so as to give rise to somewhat thinner bundles of approximately the same length as the original.

Molecular Weight and Molecular Weight Distribution. Modifications in average molcular weight and molecular weight distribution were a consequence of the mixing/processing/molding cycle of the asbestos reinforced PVC plastics. These modifications were more apparent in the high molecular weight matrix material than in the low. In general, they were not markedly different when the plastics were subjected to the processing/molding cycle without the inclusion of the asbestos.

As a rule, the number average molecular weight increased, and the higher molecular weight averages increased. This of course resulted in a significant increase in the polydispersity or the broadness of the molecular weight distribution. Finally, although further confirmation is called for, the results also suggest that the No. 1 fiber (described as clean) minimizes the changes in molecular weight and molecular weight distribution undergone by the polymers (*see* Tables V and VI).

Table VII. Glass *vs.* Asbestos-Reinforced Rigid PVC

Property	*Glass Reinforced 20%*	*Asbestos Reinforced 15%*	*Asbestos Reinforced 30%*
Impact strength, ft-lbs/in-notch	1.0–1.6	1.2	1.4
Heat deflection temperature, °F	170–180	176	176
Density, grams/cm^3	1.49–1.58	—	1.62
Flammability	non-burning	non-burning	
Tensile strength, psi	10,000	7,500	7,500
Ultimate elongation, %	1.5–4.0	2.0–1.5	
Rockwell hardness, M	M80–88	M71–93	
Cost of fiber	55–60¢/lb	4.5¢/lb	

Changes in Chemical Structure. The infrared spectra (Figure 6) showed several changes in the polymer/stabilizer combination as a consequence of processing. One of these, the loss of an absorption at 1550 cm^{-1} was tied to alterations in the stabilizer molecule. Another change involved the appearance of a band at 1030 cm^{-1}. This would suggest reactions leading to the oxidation of the S in the sulfur containing organotin stabilizer. Plausible reactions (*2*) are:

$$\mathrm{R\,S\,Sn} \xrightarrow{\mathrm{O}} \mathrm{R\,SO_3H}$$

$$\mathrm{Sn\,S\,R} + \underset{\displaystyle |\atop \displaystyle \mathrm{Cl}}{-\mathrm{C}-} \rightarrow \mathrm{SnCl} + \underset{\displaystyle |\atop \displaystyle \mathrm{S} \atop \displaystyle | \atop \displaystyle \mathrm{R}}{-\mathrm{C}-} \xrightarrow{\mathrm{O}} \underset{\displaystyle |\atop \displaystyle \mathrm{O{=}S{=}O} \atop \displaystyle | \atop \displaystyle \mathrm{R}}{-\mathrm{C}-}$$

The properties obtained in this study were compared with those in the literature for glass reinforced rigid PVC (*3*) (*see* Table VII). The mechanical and thermal properties obtained by reinforcing with the 4.5¢ asbestos are fully comparable with those obtained with chopped glass fibers.

Literature Cited

1. Cameron, A. B., Heron, G. F., Wicker, G. L., "Asbestos Reinforced Thermoplastics," *28th Ann. Tech. Conf. Reinforced Plastics/Composites,* Institute S.P.I. 1973 section 11-B, p. 1.
2. Deanin, R. D., Lowell Technological Institute, private communication.
3. Owens-Corning Fiberglass Corp., Compound Selector, Jan. 1970.

RECEIVED October 11, 1973.

5

Mica as a Reinforcement for Plastics

P. D. SHEPHERD, F. J. GOLEMBA, and F. W. MAINE

Fiberglas Canada, Ltd., Guelph, Ontario, Canada

Theoretically, mica flakes (100-μ diameter) and other platelets can reinforce plastics as efficiently as fibers for unidirectional composites; for planar isotropic composites platelets are more efficient than fibers. Strength and modulus of mica composites are a function of flake aspect ratio and volume fraction. Strengths measured to date are lower than possible, owing to flaws in the flakes. Suitably prepared mica will reinforce ABS, SAN, and nylon 6/6 to yield useful injection moldable compounds with moduli higher than present RTP's, flexural strengths comparable with RTP's, and impact strengths less than RTP's. In ternary composites of mica/glass fiber/thermoplastic resin, up to 60 wt % solids, maximum in modulus, flexural strength, and impact strength does not occur at a single composition. Therefore, in any application, desired properties must be compromised.

Flake or platelet minerals are materials not generally recognized as reinforcing elements. We have found that mica, when suitably prepared, will increase the strength and modulus of some common polymers. This paper outlines the principles of platelet reinforcement and compares the theoretical behavior of platelets as reinforcement with the behavior of fibers and spheres. Experimental results compare polymers reinforced with mica and mica/glass-fiber combinations to glass-fiber reinforced compounds.

Theoretical Principles

Polymers reinforced with platelets can be treated in a manner similar to discontinuous-fiber reinforcement. Consider a platelet composite in which the platelets are:

(1) square with side L, thickness t
(2) perfectly aligned

(3) perfectly bonded to the matrix
(4) evenly spaced
(5) linearly elastic to failure

Since the platelets and matrix have different elastic moduli, straining in tension of such a composite causes shear stress at the platelet–matrix interface. This is analogous to the case of fibers in a plastic matrix. These shear stresses transmit the applied load to the platelets, causing tensile stress, σ_p, in the platelets. Since the shear stresses increase with distance from the center of the flake (as for fibers), the tensile stress in the platelet is not constant but is a maximum at the flake center and decreases with distance from the center. Thus, there will be a critical length of platelet which must be exceeded for the flakes to be stressed to their failure stress, σ_p^*. This critical length divided by thickness is the critical aspect ratio.

Matrices that are elastic to failure have been examined by Padawer and Beecher (*1*) while ductile matrices have been considered by Shepherd (*2*). When the matrix is deforming elastically, the composite modulus (E_c) has been shown (*1*) to be:

$$E_c = E_p V_p \,(\mathrm{MRF}) + E_m \,(1 - V_p) \qquad (1)$$

where:

E_p = platelet elastic modulus
= 25×10^6 psi for mica
V_p = platelet volume fraction
E_m = matrix elastic modulus
MRF = modulus reduction factor

$$= 1 - \frac{\tanh u}{u}$$

$$u = \alpha \frac{(G_m + V_p)^{1/2}}{E_p\,(1 - V_p)}$$

α = aspect ratio
= L/t
G_m = matrix shear modulus

The theoretical moduli for mica composites have been calculated assuming $G_m = 200{,}000$ psi and are illustrated in Figure 1. The theoretical modulus for a system of fibers and spheres with the same modulus can also be calculated using Kelly's (*3*) equations for fibers and Nielsen's (*4*) equations for spheres. The results of these calculations are also shown

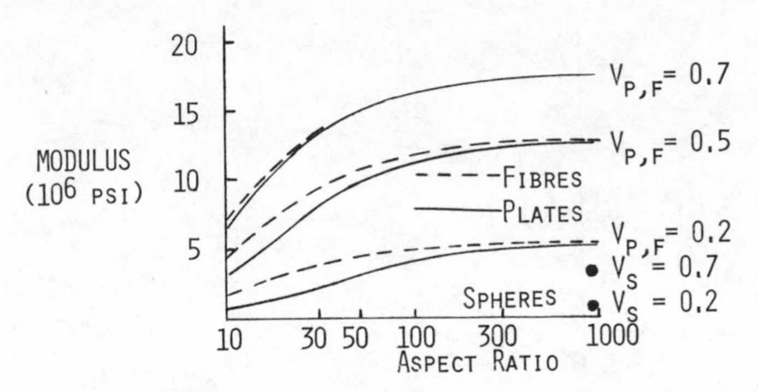

Figure 1. Theoretical modulus of mica-reinforced plastics

in Figure 1 for comparison. It is apparent that platelets and fibers are more efficient stiffening agents than spheres; for aspect ratios greater than 50, plates and fibers are essentially equivalent.

As the composite deforms, the shear stress in the matrix, near the matrix–platelet interface, and the tensile stress in the platelet increase. For matrices that are elastic to failure, the composite fails when the matrix shear strength, σ_m, is reached or the flakes fail. Padawer and Beecher have calculated the composite stress for these two conditions:

Composite stress for flake fracture =

$$\sigma_c^p = \sigma_p^* \,(\mathrm{SRF})\, V_p + \sigma_m' \,(1 - V_p) \qquad (2)$$

where:

σ_p^* = platelet tensile strength

SRF = strength reduction factor

$$= 1 - \frac{\tanh u}{u}\,(1 - \operatorname{sech} u)$$

σ_m' = stress in the matrix at composite failure

$$\simeq \sigma_m^*/3$$

σ_m^* = matrix tensile strength

Composite stress at matrix failure =

$$\sigma_c^m = \tau_m \,(\mathrm{MPF}) + \sigma_m'' \,(1 - V_p) \qquad (3)$$

where: MPF = matrix performance factor

$$= V_p \frac{\sigma}{u}\left(\frac{1}{\tanh u} - \frac{1}{u}\right)$$

σ_m'' = stress in the matrix at composite failure

$$\simeq \sigma_m^*$$

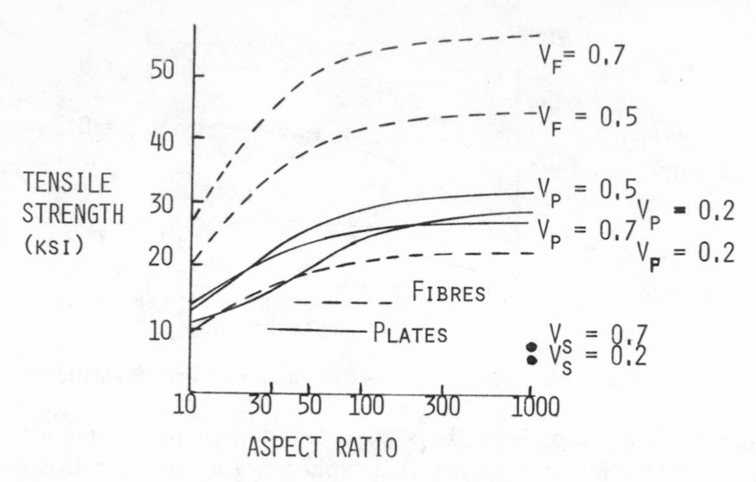

Figure 2. Theoretical tensile strength of mica-reinforced brittle plastics

The composite strength is then the lesser of $\sigma_c{}^p$ or $\sigma_c{}^m$. The strength of SAN reinforced with mica platelets, of strength 300,000 psi, was calculated from this theory and is illustrated in Figure 2 (mica strengths up to 450,000 psi have been measured by Orowan (*5*).

Also shown in Figure 2 is the expected strength of fiber- and sphere-reinforced SAN, calculated using Piggott's (*6*) and Nielsen's (*4*) equations, respectively. Again, spheres are much less efficient than fibers or platelets, but fibers are now more efficient reinforcement for volume fractions greater than 0.2. However, since platelets reinforce in a planar dimension rather than only longitudinally, platelet composites will be more efficient in producing planar isotropic composites than fibers.

For the SAN system, the strength calculations predict that the composite fails by matrix and not flake failure under all conditions. However, for some polymers the matrix will fail at low aspect ratios and the flakes will fail at high aspect ratios; the controlling factor is the matrix shear strength.

For matrices that are ductile and flow at a constant shear stress, τ_m, composite strength again depends on aspect ratio. For aspect ratios greater than $(\sigma_p{}^*/\tau_m) - 1$ (the critical aspect ratio), the flake stress reaches $\sigma_p{}^*$; for ratios less than the critical aspect ratio, the flakes do not fail but pull out of the matrix. The predicted composite strength (*2*) under these conditions is:

$$\text{for } \alpha < \alpha_{\text{critical}} \quad \sigma_c = \frac{\tau_m}{2}(\alpha + 1)V_p + \sigma_m''(1 - V_p) \qquad (4)$$

$$\text{for } \alpha > \alpha_{\text{critical}} \quad \sigma_c = \sigma_p{}^*\left(1 - \frac{\alpha_{\text{critical}}}{2\alpha}\right)V_p + \sigma_m'(1 - V_p) \qquad (5)$$

Figure 3 illustrates the predicted composite strengths for mica in a polycarbonate matrix. The critical aspect ratio for this system is 49. For aspect ratios greater than 49, the mica platelets will fracture; for aspect ratios less than 49, the flakes do not fail but are pulled out of the matrix intact. The latter case is analogous to the matrix failure case when the matrix is brittle. In any real composite system, the theoretical strengths calculated here are reduced by voids, platelet misalignment, poor matrix–platelet adhesion, and platelet flaws.

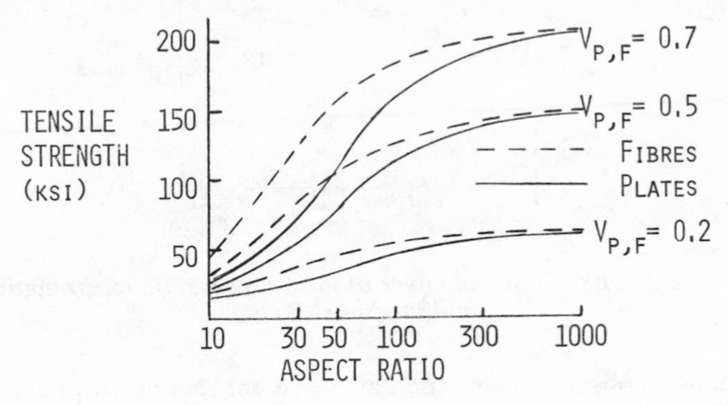

Figure 3. Theoretical tensile strength of mica-reinforced ductile plastics

The strength of an aligned discontinuous fiber composite can be calculated using Kelly and Tyson's (7) equations. For fibers of 300,000 psi strength in a polycarbonate matrix, the composite strength is also shown in Figure 3, the critical aspect ratio for fibers being 25. It is apparent that an aligned fiber composite is significantly stronger than the equivalent platelet composite for aspect ratios less than 200. However, the fiber composite is strong only in one direction; the transverse tensile strength would be about 8000 psi. The platelet composite, on the other hand, will have the same strength in the transverse direction as in the longitudinal direction. Further, if the platelets are not perfectly aligned but randomly arranged in a planar habit, the composite strength and modulus will not be a function of testing direction, as demonstrated by Economy (9).

Although it is possible theoretically to predict the strength and modulus of the three types of composites based on the previous assumptions, only the fiber- and sphere-reinforced composites can be done accurately. This is because the fibers and spheres can be characterized accurately while the platelet materials generally used cannot; for example, the statistical distribution of glass-fiber strengths is known while that for mica flakes of 100-μ diameter cannot be determined. In addition, aspect ratio measurement on a fiber is straightforward while that for mica

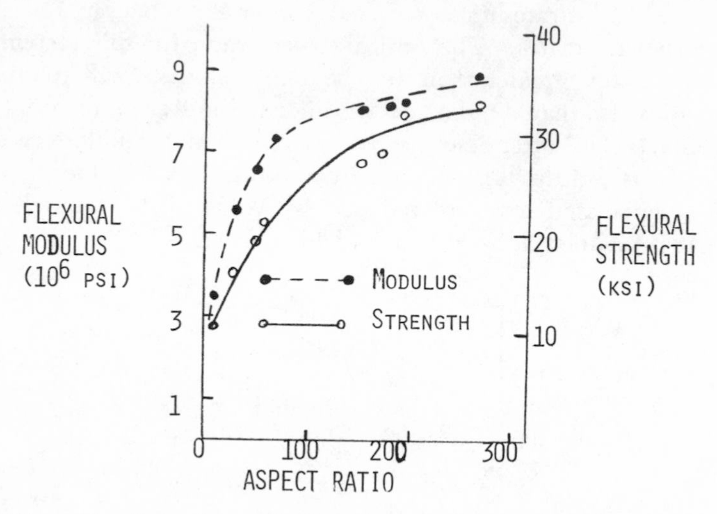

Figure 4. Experimental effect of mica aspect ratio on strength and modulus (8)

flakes cannot be absolute since the flakes are not regular in planar dimensions or in thickness. Nevertheless, experimental work has substantiated the theoretical predictions that strength and modulus are a nonlinear function with respect to aspect ratio and that strength and modulus are a linear function with respect to volume fraction; Figure 4 illustrates the dependence of strength and modulus on aspect ratio; Figure 5 illustrates the linear dependence of strength and modulus on volume fraction.

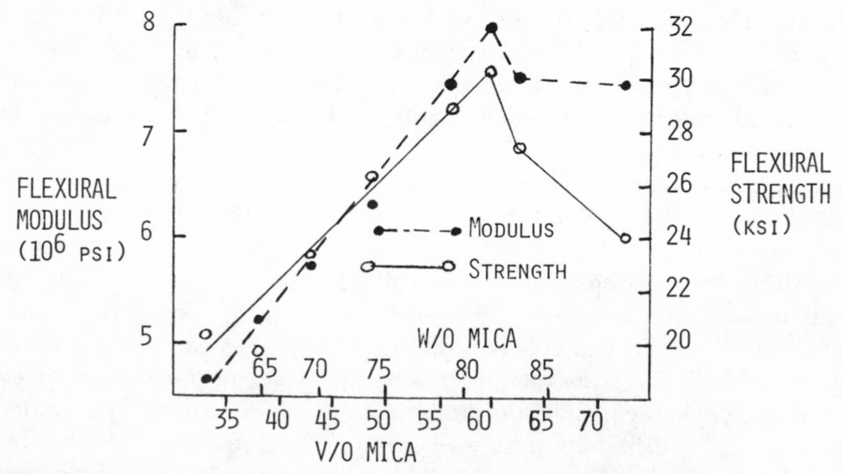

Figure 5. Experimental effect of mica volume fraction on strength and modulus

Material Properties

Although mechanical properties increase continuously up to 85 wt % mica, as shown in Figure 5, injection molding is feasible only for compositions less than 60 wt %. Two completely different types of compounds will therefore be considered: injection moldable and only compression moldable.

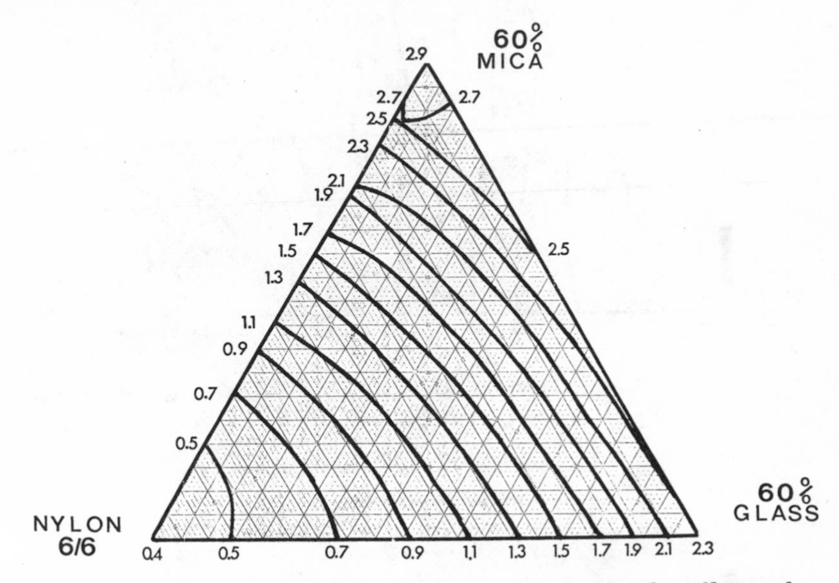

Figure 6. Flexural modulus response surface for mica/glass-fiber/nylon 6/6 composites (10^6 psi)

For the injection moldable compounds, we examined the effect of mica content and of glass/mica ratios on mechanical properties for SAN, ABS, nylon 6/6 and thermoplastic polyester matrices. The complete response surface for 0–60 wt % glass and mica has been determined for the nylon 6/6, ABS, and polyester systems. Figures 6, 7, and 8 illustrate the type of surface generated for the nylon 6/6 system for flexural modulus, flexural strength, and notched izod impact strength, respectively; the ABS, SAN, and polyester systems yield similar surfaces. These figures indicate that the maximum in modulus, flexural strength, and impact strength does not occur at the same composition. Figure 6 shows the maximum modulus to be at 0 wt % glass, 60 wt % mica, and the minimum at 0 wt % glass, 0 wt % mica; Figure 7, the maximum flexural strength at 60 wt % glass, 0 wt % mica and the minimum at 0 wt % glass, 12 wt % mica (0.2 × 60); Figure 8, the maximum impact strength at 45 wt % glass (0.75 × 60), 0 wt % mica and the minimum at 7 wt % glass, 0 wt % mica or 0 wt % glass, 60 wt % mica. Thus, no single composition gives

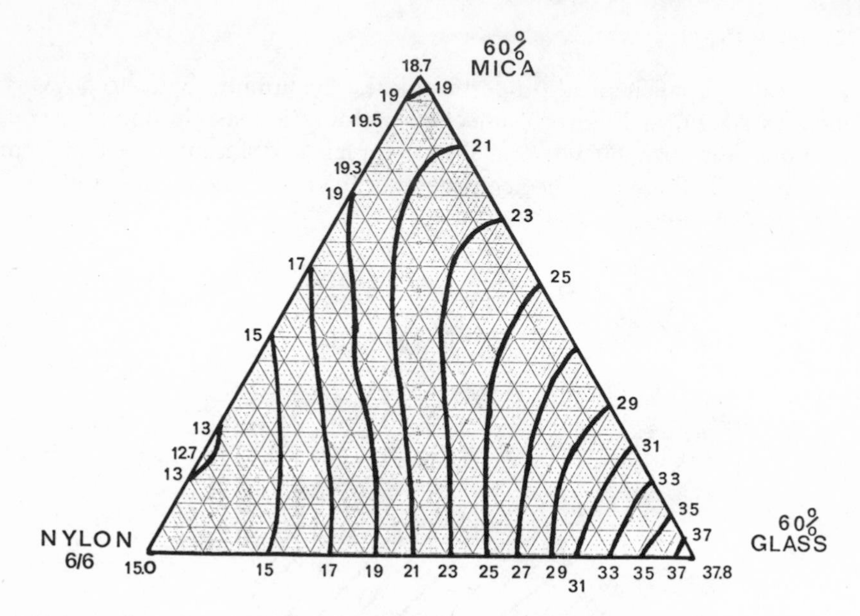

Figure 7. Flexural strength response surface for mica/glass-fiber/nylon 6/6 composites (10^3 psi)

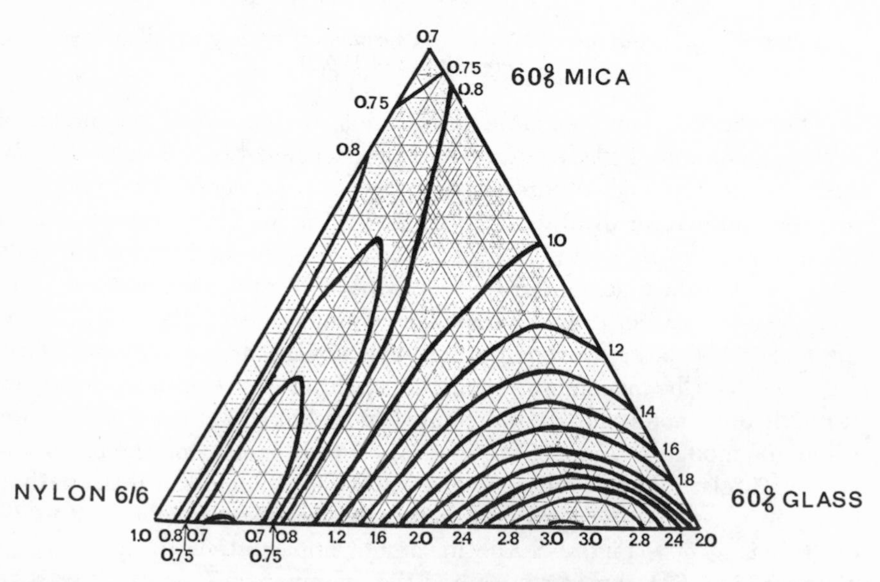

Figure 8. Notched izod impact strength response surface for mica/glass-fiber/nylon 6/6 composites (ft-lb/inch)

a maximum in all properties, and a composition must be selected to yield the desired trade-off in properties.

Two types of products would be of interest relative to existing glass-fiber RTP's: (1) high modulus, and (2) improved modulus with little loss of impact strength. For the present, we will compare the properties of six new compounds based on this work:

(1) nylon 6/6, 20 wt % glass fiber, 30 wt % mica (RFM 2030)
(2) nylon 6/6, 50 wt % mica (RM 5000)
(3) ABS, 20 wt % glass fiber, 30 wt % mica (AFM 2030)
(4) ABS, 50 wt % mica (AM 5000)
(5) SAN, 20 wt % glass, 30 wt % mica (BFM 2030)
(6) SAN, 50 wt % mica (BM 5000)

Even at these high reinforcement levels injection molding was feasible. In fact these composites molded as readily as their 20 wt % glass-fiber counterparts with cycle times equal to or less than the 20 wt % glass compound, as the data in Table I illustrate. Mechanical properties of

Table I. Cycle Times for Various Compounds

Compound	*Part Size, oz*	*Cycle Time, secs.*
20 wt % glass-fiber SAN	1.20	29
BFM 2030	1.55	23
BM 5000	1.60	22
20 wt % glass-fiber/nylon 6/6	1.45	13
RFM 2030	1.80	10
RM 5000	1.80	10
20 wt % glass-fiber/ABS	1.17	20
AFM 2030	1.46	21
AM 5000	1.50	20

the glass compounds were compared therefore with the 20 wt % glass-fiber composites compounded and molded on the same equipment. Table II summarizes the properties of the composites measured to date. In general, these new composites have higher moduli, equal strengths and heat distortion temperatures, and somewhat lower impact strengths than the comparable glass reinforced compound. The one exception is RF 2030 where all properties are equal to or greater than 20 wt % glass-reinforced nylon 6/6.

For composites greater than 60 wt % mica, only compression molding produces good composites. We have studied thermosetting and thermoplastic matrices reinforced with 50 vol % mica. Moduli much higher than existing compounds were obtained; the strengths were at least com-

Table II. Properties of ABS,

	ABS		
	20% GF	*AFM 2030*	*AM 5000*
Specific gravity	1.19	1.52	1.52
Specific volume, in^3/lb	23.4	18.3	18.3
Mold shrinkage (1/4″ section), in/in	.0006	.007	.0014
Tensile strength, psi	10,700	11,620	9,470
Tensile modulus, 10^6 psi	0.81	2.15	2.12
Flexural strength, psi	14,770	17,860	14,000
Flexural modulus, 10^6 psi	0.81	2.04	2.13
Shear strength, psi	6,220	8,410	7,410
Compressive strength, psi	11,700	17,500	16,400
Izod impact strength			
notched, ft-lb/in	1.3	0.8	0.6
unnotched ft-lb/in	4.3	2.8	2.4
Heat distortion temperature			
at 66 psi, °F	212	230	228
at 264 psi, °F	204	222	219

parable with and often greater than existing compounds. The properties of mica-reinforced thermosets studied to date are compared with existing compounds in Table III. For the phenolic compounds, the molding cycle was 3 min or equal to that for commercial compounds.

Table III. Compression-Molded Thermoset Composites (50 vol % mica)

Compound	*Flexural Strength, 10^3 psi*	*Flexural Modulus, 10^6 psi*
Experimental mica/epoxy	24.0	6.4
Experimental mica/polyester	23.0	6.8
Experimental mica/phenolic	21.0	7.5
Commercial mica/phenolic	8–10	2.5–5.0
Commercial B.M.C.	10–20	1.4–2.0

Table IV. Compression-Molded Mica Thermoplastic Composites[a]

Matrix	*Flexural Strength, 10^3 psi*		*Flexural Modulus, 10^6 psi*	
	50 vol % mica	*40 wt % glass*	*50 vol % mica*	*40 wt % glass*
Polystyrene	24.0	17.5	6.5	1.50
SAN	30.0	23.2	7.7	1.85
Nylon 6/6	27.0	42.0	6.5	1.60
Polyester	27.0	34.0	6.9	1.60
Polypropylene	25.0	10.5	5.5	0.95
Polyethylene	18.0	14.0	4.5	1.10

[a] By comparison with 40 wt % glass compounds.

SAN, and Nylon 6/6 Compounds

SAN			*Nylon 6/6*		
20% GF	*BFM 2030*	*BM 5000*	*20% GF*	*RFM 2030*	*RM 5000*
1.24	1.53	1.55	1.24	1.60	1.55
22.4	18.2	17.9	22.4	17.3	17.9
.0005	.0014	.0015	.005	.0038	.0054
14,770	12,000	9,900	15,490	17,990	13,760
1.32	2.34	2.55	.77	2.33	2.57
20,100	19,200	14,500	23,370	26,140	18,000
1.28	2.43	2.68	.94	2.04	1.93
8,870	8,600	7,100	10,320	11,050	9,510
19,000	20,270	15,560	18,900	20,200	16,000
.9	0.8	0.5	0.9	1.4	0.9
3.4	2.3	2.3	7.7	7.6	4.2
211	237	229	485	550	487
207	230	227	452	473	445

The properties of mica-reinforced thermoplastics studied to date are compared with 40 wt % glass compounds (the highest available) in Table IV. Since these compounds must be removed from the mold cold, molding times are considerably longer than the thermoset cycle, generally about 30–45 minutes. The notched impact strength of the thermoset and thermoplastic composites are all same, for practical purposes, having a value about 1 ft-lb/inch.

Conclusions

Flake reinforcement of plastics offers a unique solution to the anisotropy of fiber-reinforced plastics. Mica platelets will reinforce many polymers, both thermoset and thermoplastic, to give high modulus composites. Combinations of mica and glass-fiber with thermoplastic resins give composites of improved impact strength compared with mica-only composites.

Literature Cited

1. Padawer, G. E., Beecher, N., *Polym. Eng. Sci.* (1970) **10,** 185.
2. Shepherd, P. D., Ph.D. Thesis, University of Toronto, Canada, 1969.
3. Kelly, A., "Strong Solids," pp. 121-125, Clarendon Press, Oxford, 1968.
4. Nielsen, L. E., *J. Appl. Polym. Sci.* (1966) **10,** 97.
5. Orowan, V. E., *Z. Phys.* (1933) **82,** 235.
6. Piggott, M. R., *Acta Met.* (1966) **14,** 1429.
7. Kelley, A., Tyson, W. R., *J. Mech. Phys. Solids* (1969) **13,** 329.
8. Lusis, J., Woodhams, R. T., Xanthos, M., *Polym. Eng. Sci.* (1973) **13,** 139.
9. Economy, J., Wohrer, L. C., Matkovich, V. I., *SAMPE J.* (Dec./Jan. 1969).

RECEIVED October 11, 1973.

6

Low Quartz Microforms and Their Contributions to Plastics and Resinous Systems

JAMES E. MORELAND

Malvern Minerals Co., P.O. Box 1246, Hot Springs, Ark. 71901

Information about the study and practical use of low quartz microforms is meager. This is especially true in the applied fields of plastics composites and advanced coatings. Until recently, naturally occurring particulate minerals were considered mostly for their cost reducing contributions. These particulate forms, such as low quartz microforms, have much more to offer, such as surface energies and functionalities. Particle shape and measurement also contribute to rheological behavior. Hardness and softness, age and condition of surface, general and ionic impurity, and thermal property influence end results in both macro- and micromechanics. This article is a precursor to more definitive future papers on new and exciting developments in reinforced plastics and coatings.

This paper provides some facts on the use of finely divided naturally occurring mineral, quartz. The discussion involves geology, mining, processing, and fine particle interpretation and measurement. This mineral in micron size particulate form is low or alpha quartz or low quartz microforms. The surface chemistry of these microforms is stirring much interest in coatings and plastics.

Low Quartz Microforms

What are low quartz microforms? By 1965, 22 (*1*) distinct and separate phases of SiO_2 had been confirmed, excluding silicates (this paper deals with silicic chemistry rather than siliceous chemistry). Low quartz or alpha quartz (beta quartz in Germany) is the prevalent form in quartz rocks (Figure 1), sandstones, sands; a tripolitic form known as the lower portion of the upper novaculite is found in the novaculite uplift in western Arkansas. Low quartz microforms involve particles as small as 0.25 μ to a top size (arbitrary) of 44.0 μ (325 mesh). Intermediate particles require proper machinery and quality control. A number of graded products can be manufactured within these parameters with both wide and narrow particle size distributions.

Quartz is the hardest common mineral and resists breaking down. When it is crushed, it disintegrates (2). The highly disturbed surface of pulverized quartz is a subject in itself which is covered somewhat in this paper. We refer to freshly fractured surfaces later.

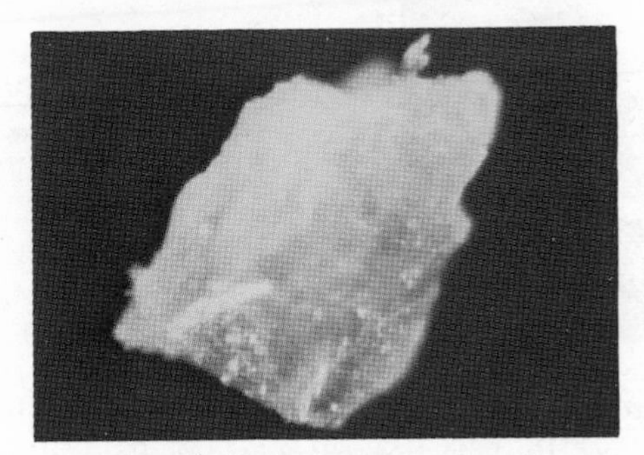

Figure 1. Quartzite particle

General

Micron-sized, naturally occurring silica, quartz, or tripoli (3) all occur in a range of 99% SiO_2 and the alpha phase of quartz. These forms are stable in their phases from absolute 0° to 573°C. As silica they reach a very viscous fluid stage at 1450°–1750°C. In attaining the melt temperature they invert first to beta quartz, then tridymite, and finally cristobalite. The inversions are reversible under controlled conditions.

The SiO_4 tetrahedra in alpha quartz are tightly packed in their crystal structure. Because of this the mineral is usually pure, and other constituents and trace elements are usually restricted to low partial percentages. The hardness of quartz is an advantage and a disadvantage. In plastic composites the hardness provides both mechanical strength and abrasion resistance. The hardness causes the mineral to be of high modulus, and in low modulus resins the result is brittleness. Some brittleness can be eliminated by silane treatment or straight admixing of silane. The characteristics discussed here provide coatings and plastics with an economical extender/filler which is thermally stable, pure, low in ionic impurities, and hard. Other more refined characteristics are covered below.

Microform Technology

Although this paper is a general study of several low quartz microforms, it deals largely with the knowledge and technology involving microcrystalline novaculite. The ultimate particulate is microcrystalline granular, lameller (Figure 2). The particle which is magnified 208,000 × is highly unusual for quartz because of its platey structure. These particles come from anhedral clusters (*4*) and are low quartz in a bound state of subdivision (Figure 3) (*cf.*, a quartz sand particle in Figure 1). The

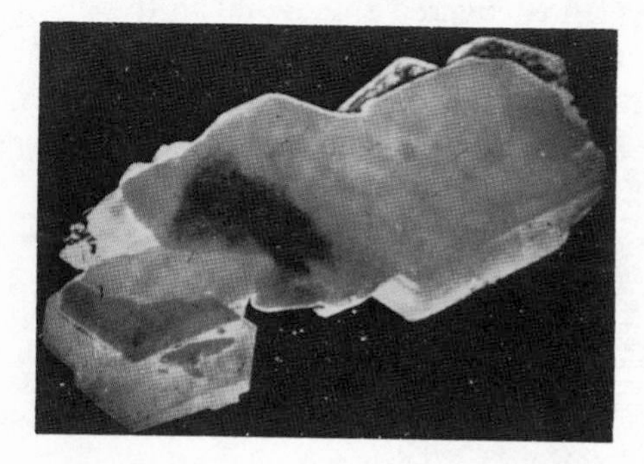

Figure 2. Upper novaculite particulate

particulate (Figure 2) would indicate the free state of subdivision. Quartz in these forms offers the plastics or coatings field distinct advantages such as lower abrasiveness, lower oil adsorption, and distinct surface characteristics. In a product from microcrystalline novaculite, particles larger than 3–7 microns (*5*) would be considered a cluster and would be quite friable. One might imagine, for instance a 44-μ particle (325 mesh) which could be composed of approximately 3000 individual particulates, assuming that the particulate would measure $5\ \mu \times 3\ \mu \times 1.5\ \mu$. Computation is based on cubic particulates and clusters. The relatively low abrasiveness of these microforms of novaculite is caused by the lack of the usual jagged and sharp points found in the pulverized varieties of silica.

Low oil adsorption, offhand, would seem to be the result of lower surface area. However, surface measurement shows microcrystalline novaculite with a greater surface area than ground silicas, even though the latter indicate lower average particle size and a wealth of particles measuring less than 1μ. Since the novaculite particles are lameller and platey, the interstitial surface area must come into play. We cannot discuss thoroughly the differences between tripolitic novaculite (tripoli) and the pulverized silicas (*6*) but we can observe a type which is unground or unpulverized. Geologically this type will have an aged surface functionality and be relatively undisturbed. The freshly fractured or pulverized silica surfaces are described as highly disturbed (*7*), carrying

Figure 3. Upper novaculite anhedral cluster

with them a layer of nonquartz silica 20–60 mμ thick. The undisturbed type of surface from the natural microform is helpful in keeping mix viscosities low in measurement, but thickening effects of the crushed varieties of quartz have definite advantages in extending elastomers and where favorable thixotropic behavior is important. No single product can answer, all needs, and the usual tradeoff of properties arises. However, difference in particle size shape, range of distribution of particle size, surface characteristics, and origin of crystal structure are important. The effects can cause dramatic differences in performance.

Table I. Effect of Union Carbide A-1100 Silane Coupling Agent on Nylon 6[a] and Nylon 6.6[b] Filled with Novacite L-207A

—Courtesy Union Carbide Corp.

	Nylon 6.6/ 50% Novacite L-207A			*Nylon 6/ 50% Novacite L-207A*		
	Unfilled	*No Silane*	*A-1100*	*Unfilled*	*No Silane*	*A-1100*
Flexural Strength, psi						
Dry	16,800	10,500	16,500	12,500	15,500	17,700
16 hrs/50°C H_2O	9,800	6,000	13,400	6,700	7,400	10,100
7 days/50°C H_2O	5,800	4,800	9,700	5,100	5,300	7,200
Flexural modulus, 10^5 psi						
Dry	4.2	9.2	9.4	2.7	8.0	7.1
16 hrs/50°C H_2O	1.7	3.4	4.7	0.8	2.7	2.3
7 days/50°C H_2O	1.0	1.4	2.2	0.8	1.0	1.3
Tensile strength, psi						
Dry	11,200	4,600	10,100	9,000	8,300	9,300
16 hrs/50°C H_2O	7,700	3,300	7,400	6,700	4,600	6,700
Dart impact strength, inch-lbs.	14	<4.5	<4.5	600	<4.5	17

[a] Plaskon 8201, Allied Chemical Co.
[b] Zytel 101, Dupont.

Surface Functionality and Coupling Agents

Angewandte Chemie (June 1966) reported, "The three dimensional periodic arrangement of atoms in a crystalline solid is interrupted at a surface. . . . The surface may therefore be regarded as an extreme lattice defect." For quartz, visually we can imagine how a cell might look:

The larger circles are oxygen atoms while the smaller dark circles are silicon atoms. Superimposed at the top of the cell are the hydrogen atoms at what is shown to be a cleavage or fractured area. At this terminal is the silanol surface. Euhedral faces are also terminated in SiOH, as is glass.

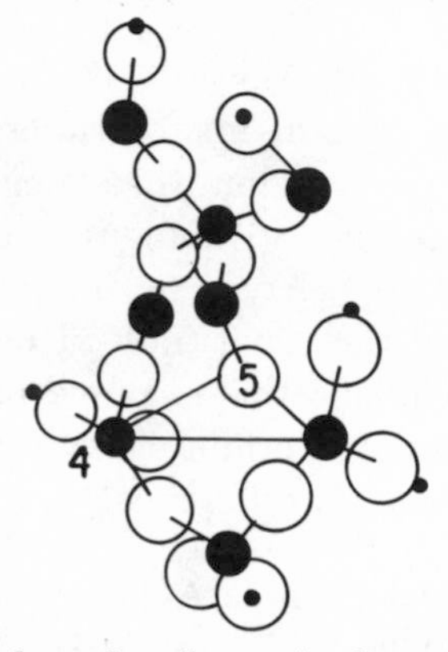

The availability of the silanols is the basis for a commercial family of products. These are silane-treated quartz microforms (novaculite) (*8*). No other silane-treated naturally occurring silicas are commercially available in the United States. Some clays (*9*) are available with silane treatment, but, novaculite and clay are treated quite differently.

Table II. Use of Novacite and Novakup in Polypropylene (Injected Molded)

—Courtesy Dow Corning Corp.

Product	*ASTM Flex, psi*	*ASTM Tensile, psi*	*ASTM Flex Mod., psi*
Control (Novacite)	6610	3227	315,000
Novakup 6020 0.5	7547	3790	310,000
Novakup 5069 0.5	8150	4249	310,000

By approximate calculation of the silanol population, surface area, and particles per gram, an effective concentration of coupling agents can ordinarily be attached to the quartz surface. It is the ambifunctionality by crosslinking across the particle face to a plastic *via* an organofunctional group, where the real interests lie. Several groups can react with the SiOH—*e.g.*, ethoxy, methoxy, or acetoxy. The organofunctional group could be epoxy, amine, polyamino, methacrylate, mercapto, styrylamine, and so forth. New families of reinforced, mineral filled plastics and coatings are appearing—*e.g.*, mineral reinforced nylons, polypropylene, DAP, and epoxy—all of which are presented in Tables I–IV. Not shown, but very active, are phenolic tank linings, coil coatings, polyester adhesives, and resistor varnishes. Since the novaculite microform can be readily treated in substantial quantities and can support all ambifunctional silanes, the interest in these products is broad.

Tables I–IV indicate the effectiveness, or lack of it, when small concentrations of different silanes are used in several plastics. Variations in silane concentration, further study into characteristics of minerals, consideration of the isoelectric point (ZPC) of the mineral surface, and new silanes will all change the role of the minerals from filler to reinforcing agent. At the same time the ability of manufacturers to develop machinery which will process and cut plastics with mohs hardness greater than 6.5 will be needed. Both manufacturers of plastics and raw material suppliers will be affected by the need and supply of improved machinery.

Mining and Processing

The text below describes some of the techniques in mining, ore selection, ore stockpiling, processing, particle size classification, and distribution of the quartz microform. There is a geologic uplift covering about 2700 square miles stretching on a west-southwest axis from just west of Little Rock, Ark., continuing to a point just east of the Oklahoma border (8). It is called The Novaculite Uplift, from novacula, a Latin word meaning razor-like. This razor reference more aptly describes conchoidial fractures coming from quartz-flint rocks. There are three strata of novaculite—lower, middle, and upper.

The lower novaculite is a cherty like stone, the source of the famed hard Arkansas stone . The rare, blue-white variety of this stone is becoming a collector's item and is almost impossible to find unfractured. The middle novaculite is usually a mixture but is highly silicic. Some of the middle novaculite in an ore body is as black as graphite but contains over 95% SiO_2.

The upper novaculite is not compacted, and in one deposit it is a very pure, grey-white powdery deposit. It is overturned, and since it is inverted, its high purity has been preserved for about 350,000,000 years.

Figure 2 shows the particulate and Figure 3 the cluster. If the ultimate particulate is less than 5 microns and the larger clusters are loose agglomerates, it should not be difficult to imagine upper novaculite ore as highly friable and hence very powdery when it is ready for processing. The ore is typical—60% minus 200 mesh and 43% minus 325 mesh.

Table III. Electrical Properties of Novacite

Type	*(Control) as Received Flex, psi*	*2 hr Boil Flex, psi*	*72 hr Boil Flex, psi*	*DK(Dry) $10^2/10^5$*
1250 Novacite (control)				4.14/3.87
1250 Novakup 6040 0.5[c]				3.24/3.08
1250 Novacite 6050 0.5[d]				3.00/2.85
1250 Novacite 1100 0.5[e]				3.01/2.86
L-207 A Novacite (control)	17,100	14,500	11,300	4.25/3.96
L-207 A Novakup 6040 0.01[f]	19,400	17,100	14,000	4.17/3.89
L-207 A Novakup 6020 0.1	19,900	20,500	18,500	4.22/3.96

[a] 100 pts. Dow Epoxy D.E.R. 331, 18 pts. Ca Z, 50 pts. Novacite.
[b] Wet—24 hrs. in 23°C water.
[c] Epoxysilane.
[d] Polyaminosilane.

There is some rock (seams) in the crude which is necessary for processing flow. However, the job is mostly 90% air floating and 10% regrind on tailings. The ore has a maximum free water take up (adsorbed) of about 10%. This value is close to a reported one (*10*) of 8.7 grams H_2O/100 grams novaculite to saturation.

Classification of product comes from separators, dust collectors, screens (coarser grades 100 mesh and coarser), and super classification (grades less than 15μ). Quality control means particle size control,

Table IV. ASTM Properties for Novacite and

Type	*Heat Deflection, °C*	*Rockwell Hardness, M*	*Water Absorption, %*	*Flexural Strength, psi*
1250 Novacite (control)	156	106	0.16	9,500
1250 Novakup 6075 1.0[b]	151	109	0.13	13,000
1250 Novakup 6031 1.0[c]	151	108	0.16	12,000

[a] 50 Pts. Dapon 35 resin, 1.5 pts. *tert*-butyl perbenzoate, 1 pt. calcium stearate, 50 pts. Novacite.

and Novakup in Epoxy Composites[a]

—Courtesy Dow Corning Corp.

DK(wet)[b] *10²/10⁵*	*DF(dry) 10²/10⁵*	*DF(wet) 10²/10⁵*	*Volume Resistivity 10²/10⁵ CPS ohm/cm (× 10¹⁵)*		*Flexural Modulus (Dry)*	*Flexural Modulus (Wet)*[g]
			Dry	Wet		
4.31/3.94	.0060/ .0271	.0153/ .0296	3.2	2.4	.695 × 10⁶	.623 × 10⁶
3.29/3.10	.0056/ .0191	.0062/ .0204	1.9	2.8	.645 × 10⁶	.662 × 10⁶
3.08/2.89	.0057/ .0180	.0070/ .0188	3.6	3.6	(1250 Novakup 6040 1.0)	
3.04/2.89	.0076/ .0179	.0081/ .0182	1.6	1.3		
4.29/4.01	.0055/ .0284	.0057/ .0280				
4.24/3.94	.0056/ .028	.0058/ .029				
4.32/4.02	.0054/ .026	.0053/ .029				

[e] Aminosilane.
[f] Epoxysilane.
[g] 4 Hr. boil.

elimination and control of contamination, and color. Particle size control is guarded with Fisher sub sieve sizer, Andreason pipette, U.S. sieves, Buckbee-Mears micromesh screens, and microscopic measurements. Contamination is controlled by diligence in manufacturing, and the particle size quality control will reveal contaminants. The color of the typical product is controlled by selective mining. Quality control is generally consistent with the novaculite microform, but variables in machinery, weather, and other factors require systematic checking, especially, if problems and complaints are to be kept at a minimum. One company

Novakup in Dapon Diallyl Phthalate Resins[a]

—Courtesy FMC Corp.

Dielectric Constant, 10³/10⁶ Hz	*Dielectric Constant (Wet)*	*Dissapation Factor 10³/10⁶ Hz*	*Dissapation Factor (Wet)*	*Insulation Resistance (Wet) after 16 hrs. @ 130°C (ohms)*
3.8/3.7	3.9/3.7	.008/.006	.019/.006	3.6×10^{14}
3.8/3.7	3.8/3.7	.007/.006	.007/.006	1.2×10^{14}
3.8/3.8	3.9/3.8	.010/.006	.017/.006	6×10^{15}

[b] Vinylsilane.
[c] Methallylate silane.

strip mines a deposit. The deposit is calculated to have approximately a million tons of ore. Overburden is a constant, severe problem because on top is the flint or cherty lower novaculite in massive thicknesses. However, the deposit of crude ore has been protected since Devonion days.

Literature Cited

1. Sosman, Robert B., "The Phases of Silica," (Preface), Rutgers Press, New Brunswick, 1965.
2. Cady, Walter G., "Piezoelectricity," p. 410, McGraw-Hill, New York, 1946.
3. "Pigment Handbook," Vol. I, p. 129, Interscience, New York, 1973.
4. "Petrographic Analysis," Malvern Minerals Co., 1950.
5. Keith, M. L., Tuttle, O. F., "Significance of Variations in the High-Low Inversion of Quartz," *Amer. J. Sci.* (1952) (Bowen Volume) 244.
6. "Pigment Handbook," Vol. I, p. 135, Interscience, New York, 1973.
7. Iler, Ralph K., "Colloid Chemistry of Silica and Silicates," p. 259, 1955.
8. Griswold, L. S., "Arkansas Geological Survey," Vol. III, "Novaculites," p. 195, 1892.
9. "Silane Adhesion Promoters in Mineral-Filled Composites," Union Carbide Corp., 1973.
10. Ledoux & Co., Inc., Rept. No. 780169-A (Dec. 23, 1959).

RECEIVED October 11, 1973.

7

Acrylic Modification of Plasticized Poly(vinyl chloride)

JOHN T. LUTZ, JR.

Rohm and Haas Co., Research Laboratories, P.O. Box 219, Bristol, Pa. 19007

Acrylic, methacrylate butadiene styrene (MBS), and acrylonitrile butadiene styrene (ABS) modifiers are commonly used in rigid PVC to improve processing and physical properties. Although plasticized PVC is generally easy to process and properties are readily adjusted through plasticizer type and concentration, the modifiers are useful for imparting unusual combinations of properties. The acrylic and MBS impact modifiers are the more effective for improving the low temperature flexibility of semi-rigid and polymeric plasticized systems, thereby permitting the use of more permanent plasticizers and the attainment of low temperature properties normally associated with the use of the non-permanent monomeric plasticizers. By using the proper acrylic or MBS modifier, abnormally high filler contents can be used in plasticized PVC compounds without excessive loss of physical properties.

Poly(vinyl chloride) (PVC) is a versatile polymer primarily because it can be modified to produce compounds having an almost infinite number of diverse properties. Rigid (unplasticized) PVC products are the most difficult to produce because of the poor cohesion of hot PVC under shear. Since 1955, a growing family of acrylic modifiers has made processing of rigid PVC much easier and more economical (by improving homogeneity, hot strength, and elongation) and has made it possible to produce rigid products with very high impact strength.

Although the use of the acrylic modifiers in rigid PVC is now commonplace, their contributions to quality and performance in plasticized PVC are still not fully utilized. Since exposition of each of the several types of acrylic modifiers would merit a paper of its own, we will only outline the utility of each and devote the remainder of this paper to the uses of impact modifiers in plasticized PVC.

Table I. Effects of Modifier on

Material	Modifier	*Shore A Hardness*	*Low Temperature* Flexibility T_f, °C	Impact T_B, °C
Di(2-ethylhexyl) adipate[8],				
δ = 8.48[b]		77–76	−51	−58
10% Modifier	ACR[1]	79–76	−55	−61
	MBS–11[2]	75–73	−53	−57
	ABS–2[14]	75–73	−53	−57
20% Modifier	ACR	82–80	−64	−62
	MBS–11	76–76	−55	−67
	ABS–2	76–76	−55	−57
Di(2-ethylhexyl) phthalate,				
δ = 8.9		83–80	−19	−30
10% Modifier	ACR	85–82	−25	−34
	MBS–11	85–81	−25	−33
	ABS–2	84–81	−25	−33
20% Modifier	ACR	87–79	−29	−43
	MBS–11	79–76	−27	−43
	ABS–2	79–77	−27	−43
Polyester[5], δ = 9.2				
10% Modifier		90–88	−4	−11
	ACR	80–87	−5	−18
	MBS–11	87–85	−7	−22
	ABS–2	88–86	−7	−20
20% Modifier	ACR	88–85	−12	−22
	MBS–11	86–83	−12	−31
	ABS–2	86–84	−11	−18

[a] Seven days/60°C, 0.5 psi, natural foam latex rubber.
[b] Solubility parameter, after the method of Burrell and Small (*4*).

Processing Aids:
- Faster fusion, especially with polymeric plasticizers
- Greater hot strength and room temperature toughness
- Improved calender tracking and sheet quality
- Reduction or elimination of plate-out
- Stabilization of cell structure in foams

Impact Modifiers:
- Improvement in low temperature toughness without sacrificing modulus or permanence
- Improvement of flex fatigue resistance and retention of tensile properties even at high filler levels
- Improved drape and drier feel of highly plasticized PVC

Lubricant-Processing Aids:
- Good release from hot metal, but essentially non-migratory at room temperature

PVC[10, d] Containing 50 phr Plasticizer[c]

Tensile Strength, psi	Ultimate Elongation, %	100% Modulus, psi	Migration into Foam Rubber, %[a]
2500	405	1100	—
2300	375	1000	—
2300	380	1000	—
2100	310	1100	—
2200	330	900	—
2200	350	900	—
2100	320	1000	—
2600	400	1700	13.0
2600	460	1400	13.1
2700	470	1400	13.2
2500	400	1500	14.3
2500	440	1300	13.5
2600	510	1200	13.7
2400	470	1400	15.6
3100	400	2400	0.4
3100	420	2000	0.5
3100	435	2000	0.5
2700	350	2000	0.7
2900	430	1700	0.5
2900	460	1600	0.6
2700	360	1700	0.7

[c] For reference: δ PVC = 9.7; GRS = 8.1; MMA = 9.5.
[d] Superscript numbers refer to compounds listed in text.

Flatting Agents:
Good dulling of surfaces without sacrifices in physical properties or soil resistance

Reactive Monomers:
Impart low viscosity to PVC/plasticizer dispersions and yield hard, tough, heat-resistant products

Detailed performances of the processing aids and reactive monomers in plasticized PVC are presented elsewhere (*1*, *2*, *3*).

Types of Impact Modifiers

Three types of impact modifiers are available:

A. All acrylic (ACR)
B. Methacrylate butadiene styrene (MBS)
C. Acrylonitrile butadiene styrene (ABS)

Within each group, the composition and quantity of the rubbery and hard phases can be varied to regulate performance, refractive index, etc. Most of the compositional information has not been disclosed, and since the complexity of the polymers precludes meaningful calculation of solubility parameters or other expressions that would clearly define the mode of operation of the modifiers, this presentation will be a pragmatic one.

Interaction of Plasticizer Type and Modifier Type and Concentration

The data in Table I illustrate the effects of modifier type and concentration on the physical properties of PVC compounds plasticized with 50 phr (parts per hundred of resin—*i.e.*, PVC + modifier) of three plasticizers of varying solubility parameters (δ). The effect of plasticizer type is much more dramatic on the properties of the unmodified PVC than is the type of modifier. Modified compounds containing the least compatible (with PVC) plasticizer, di(2-ethylhexyl) adipate (DOA), show the least improvement in low temperature properties and the most

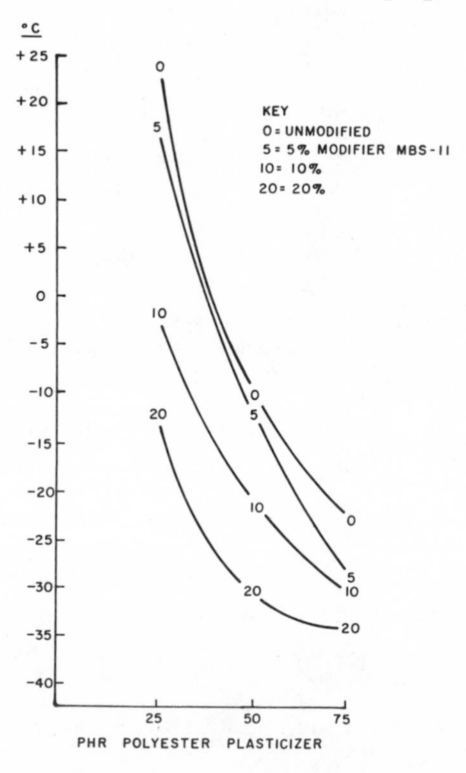

Figure 1. Brittle point

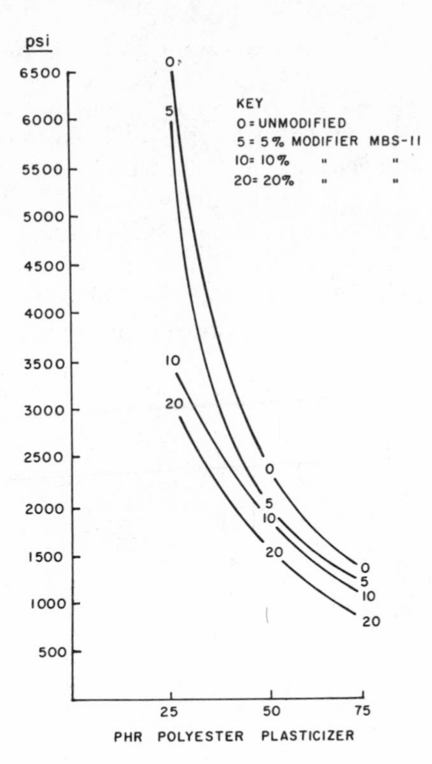

Figure 2. 100% modulus

impairment in physical properties. Generally, both the ACR and MBS modifiers are significantly better than the ABS modifiers for overall performance.

With the more compatible plasticizers, DOP and the polyesters, addition of the ACR and MBS modifiers results in greater improvements in low temperature properties (especially impact resistance) without impairment in physical properties. The addition of the ABS modifier, however, consistently results in impaired tensile strength and generally impaired elongation along with the general improvement in low temperature properties.

The greatest improvement in low temperature properties is achieved by adding MBS modifiers to PVC plasticized with polyester plasticizers. Extraction and migration resistance of the compound are not significantly impaired. Thus, two antagonistic properties—permanence and low temperature flexibility—can be imparted to a flexible compound through the use of a polyester plasticizer and an acrylic impact modifier—especially of the MBS type.

Interaction of Plasticizer Concentration and Modifier Level

PVC's containing 25, 50, and 75 phr polyester plasticizer and the MBS modifier at levels of 0, 5, 10, and 25% have been chosen as the model system. Improvement in low temperature impact resistance (brittle point, Figure 1) is not linear with increasing modifier content. The greatest percentage improvement is achieved with 10% modifier, and the greatest numerical improvement is achieved at 20% modifier (within the limits of this study). The greatest contributions of the modifier are obtained at the lowest plasticizer concentration—or, where it is needed most.

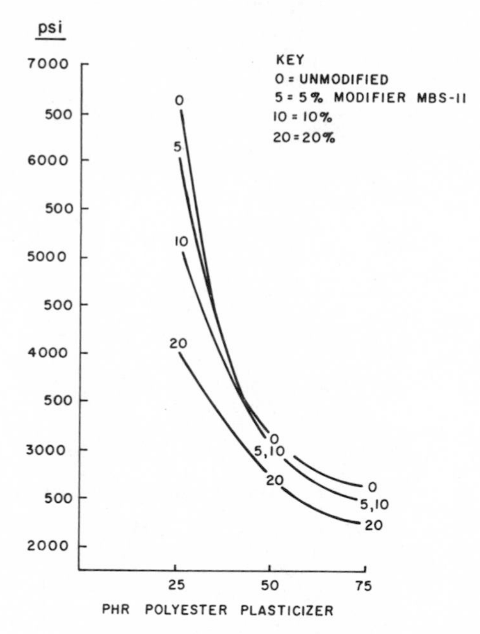

Figure 3. Tensile strength

Modulus (Figure 2) and tensile strength (Figure 3) show the most significant reductions at the lowest plasticizer concentration. Elongation (Figure 4), however, has a somewhat different response. Incorporation of 20% modifier in the 25 phr plasticizer compound results in dramatic improvement in elongation. At 50 phr plasticizer, the effects of modifier even at 20% are quite small. Significant improvements in elongation through increased modifier content are again obtained with 75 phr plasticizer. The addition of modifier results in slightly softer stocks (Shore A hardness, Figure 5), with the differences being minimized as the plasticizer concentration is increased.

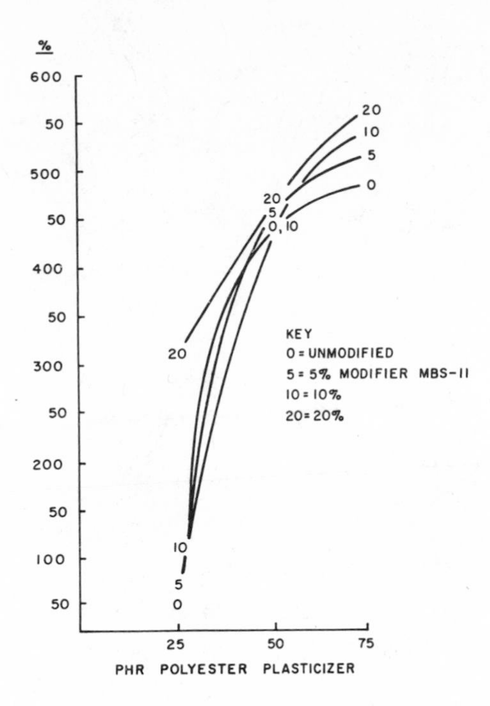

Figure 4. Ultimate elongation

Potential Applications Utilizing Modifiers

Various types of pressure-sensitive tapes and foils require the use of medium-to-high molecular weight polyester plasticizers to prevent migration from the PVC film into the adhesive mass. Frequently accompanying this requirement is the need for good low temperature impact resistance. A specific example of a product that would benefit from the use of the impact modifiers is electrical tapes (Table II). Improvements in low temperature properties were previously obtained in this type of formulation by replacing part of the polymeric plasticizer with a monomeric low temperature plasticizer such as diisodecyl adipate. The use of the acrylic impact modifiers can result in equal or greater improvement in low temperature properties without introducing migratory monomeric pasticizers that destroy adhesion.

Semi-rigid PVC sheet is being used in many applications requiring a high degree of toughness. Book covers, for example, must have good resistance to flex fatigue and be low in cost. The addition of acrylic modifiers to PVC compounds containing 20 phr DOP and 18 phr $CaCO_3$ filler gives substantial improvements in flex life (Figure 6). In this area, MBS modifiers appear to be more effective than all-acrylic (ACR) modifiers, but both are substantially more effective than ABS.

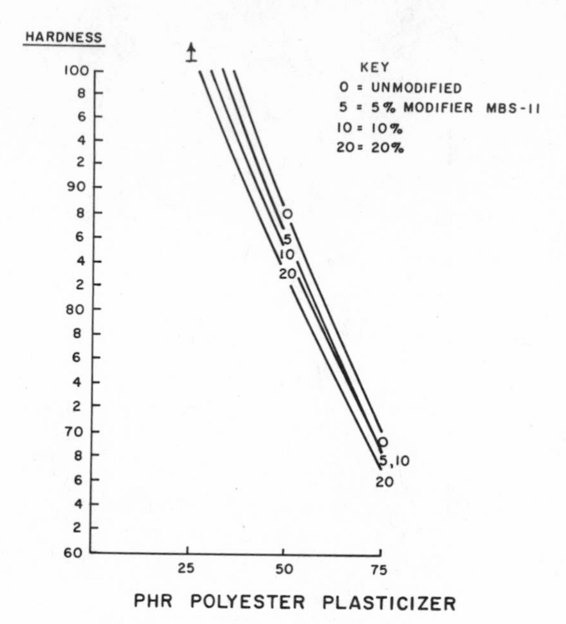

Figure 5. Shore A hardness, 10 sec

Table II. Effects of Modifiers in Electrical Tapes

	Modifier				
Property	*0%*	*ACR, 10%*	*MBS–11, 10%*	*ACR, 20%*	*MBS–11, 20%*
Shore A hardness	87–83	86–82	84–80	83–79	83–79
Tensile strength, psi	2800	2800	2800	2500	2400
Ultimate elongation, %	350	350	375	320	370
50% Modulus, psi	1100	1100	1000	900	800
100% Modulus, psi	1700	1500	1400	1300	1100
T_f, °C	−10	−10	−10	−16	−15
T_B, °C	−17	−23	−23	−26	−26
Masland impact, °C[a]	−8	−10	−12	−15	−21
Dielectric strength[a], v/mil; 0.2 in. electrodes in air:					
At 23 °C	1930	1770	1790	1840	1710
% Retention after 24 hr in distilled water at 23°C	86	78	74	67	65

[a] Determined on 7- mil film; other tests run on 75-mil panels.

Formulation: 100-X PVC[12,b]
X Modifier
55 Polyester plasticizer[6]
4 Lectro 60 (N/L Industries)
0.6 Polyethylene, AC-629
1.0 Micronex Black (ground in Paraplex G-59)

[b] Superscript numbers refer to compounds listed in text.

In addition to improvements in flex life, the use of the MBS modifiers provides a means of adjusting physical properties without changing plasticizer levels (Table III).

The use of high levels of filler is frequently desired to lower cost and to impart certain other properties such as flame retardance. The

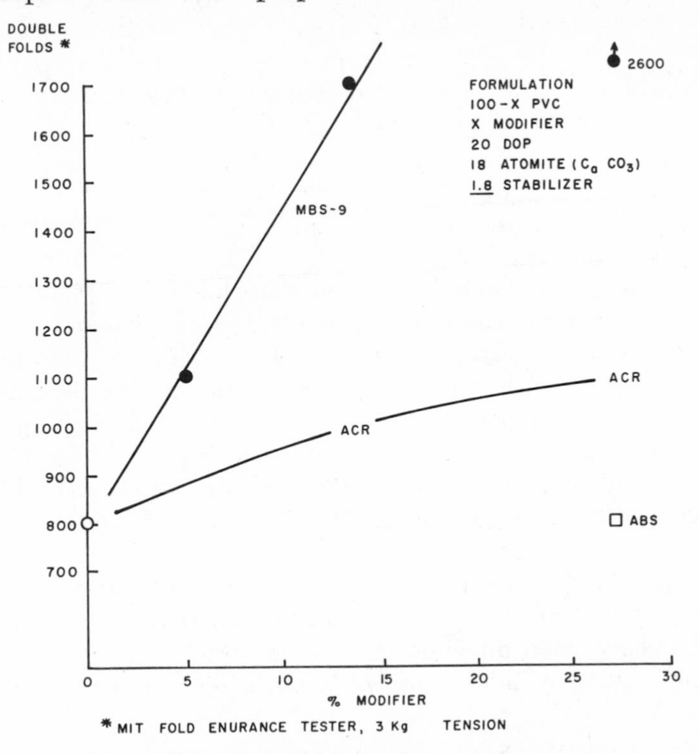

Figure 6. Effect of modifier on fold endurance

Table III. Effects of Impact Modifiers on Physical Properties of Semi-Rigid PVC[a]

	Modifier			
Property	*MBS-9*[3,b] *13.5%*	*MBS-9, 27%*	*ACR*[1], *27%*	*ABS-1*[13], *27%*
Tensile stress, psi				
Yield	4,400	3,900	3,600	4,100
Break	4,000	3,400	3,600	3,200
Elongation, %				
Yield	5	5	5	5
Break	220	200	190	155
Elastic modulus, psi	202,000	172,000	172,000	184,000

[a] Formulation: PVC[11] modifier/DOP/atomite/stabilizer = 100-X/X/20/18/1.8.
[b] Superscript numbers refer to compounds listed in text.

Table IV. Effect of Filler Level on Physical Properties

Property	*30% MBS-19 Modifier, Plasticizer 65[a], Filler[b], phr*			*30% ABS-1[13,b] Modifier, Plasticizer 60[a], Filler[b], phr*	
	0	*50*	*75*	*0*	*50*
Shore A hardness	73	76	76	76	77
Tensile strength, psi	1700	1600	1500	1500	1100
Ultimate elongation, %	375	370	355	280	230

[a] Monoplex S-75 (high molecular weight epoxidized ester)/TOTM = 1/1.
[b] Formulation: PVC[11] modifier/plasticizer/TiO_2/stabilizer/Atomite = 70/30/as indicated/5/2/as indicated.
[c] Superscript numbers refer to compounds listed in text.

acrylic resins have the ability to "bind in" fillers and pigments as evidenced by their dramatic effects on eliminating plate-out. This effect also permits one to incorporate very high levels of filler into plasticized PVC without seriously impairing tensile properties. In the comparisons in Table IV, the compound containing 30% MBS and 75 phr $CaCO_3$ has greater elongation and equivalent tensile strength and hardness when compared with the compound containing 30% ABS and no filler. The cost savings through the use of MBS over ABS become immediately obvious.

Finally, recent developments in automotive applications require compounds that are inexpensive (as raw materials and in fabrication) and that have good dimensional stability and low temperature impact resistance. Bumper filler strips or sight-shields, fender extensions, etc.

Table V. Sag Resistant-Low Temperature Impact Resistant Compounds

Compound	*Compound No.*				
	450	*451*	*452*	*453*	*455*
PVC[10,a]	100	100	90	80	80
MBS modifier[2]	0	0	10	20	20
DOP	38.7	0	0	0	0
DOA	0	30.6	22.5	22.5	0
DDA	0	0	0	0	22.5
Paraplex G–62	4.3	4.3	2.5	2.5	2.5
Stabilizer	2.3	2.3	2.3	2.3	2.3
Aluminum pigment	2.0	2.0	2.0	2.0	2.0
Shore D hardness	52–45	54–48	64–58	62–55	64–58
T_B, °C	−14	−24	−24	−47	−50
Sag index ($\times 10^{-4}$)[b]	16	10	6.3	8	5

[a] Superscript numbers refer to compounds listed in text.
[b] Sag index = inches of sag/°C/seconds to sag 3 inches.

must resist deformation at elevated temperatures in service and must survive impact over a range of temperatures.

Obvious advantages for PVC compounds are cost and the ease of injection molding or extruding parts that can be integrally colored or finished without pretreatment. Developing PVC compounds that will be stiff enough to resist sagging and still have good low temperature properties can only be accomplished, we believe, through the proper selection of plasticizer and acrylic impact modifier.

The data in Table V illustrate the possibilities of developing compounds that will have a brittle temperature of −50°C (−58°F) while being stiff enough to resist sagging in a cantilever test at temperatures up to about 40°–50°C (104°–122°F) in a 70-mil thick flat section. Conventionally plasticized and unmodified compounds having brittle temperatures of only −24°C (−12°F) literally collapse at temperatures of 32°–37°C (90°–98°F).

Conclusions

A family of acrylic modifiers for plasticized PVC is available to:

(1) Make processing easier and more economical.

(2) Impart special effects such as low gloss without sacrificing soil resistance and physical properties.

(3) Allow compounders to develop compounds having antagonistic properties such as: (a) good permanence and good low temperature flexibility; (b) sag resistance and stiffness plus excellent low temperature impact resistance; (c) stiffness and good flex fatigue resistance; (d) low cost through high filler loadings without major sacrifices in tensile properties.

We feel that we have only uncovered the tip of the iceberg when it comes to the advantages of using the acrylic modifiers. We hope that compounders will help uncover the rest through imaginative use in their work.

Materials Used

(1) Acryloid KM-323—all acrylic modifier
(2) Acryloid KM-611—MBS modifier
(3) Acryloid KM-229—MBS modifier
(4) Experimental modifier 6819-XP—MBS modifier
(5) Paraplex G-54—polyester
(6) Paraplex G-59—polyester
(7) Paraplex G-62—epoxidized soybean oil
(8) DOA—di(2-ethylhexyl) adipate
(9) DDA—Diisodecyl adipate
(10) Poly(vinyl chloride), ASTM D-1755-66, Type GP-605000
(11) Poly(vinyl chloride) ASTM D-1755-66, Type GP-405000

(12) Poly(vinyl chloride/vinyl acetate), ASTM D-2747-66T, Type C118360
(13) ABS-1, ASTM D-1778-68, Type 543
(14) ABS-2, ASTM D-1788-68, Type 611

Literature Cited

1. Lutz, J. T., Jr., Society of Plastics Engineers, Annual Technical Conference (May 1970).
2. Lutz, J. T., Jr., *Mod. Plastics* (1971) **48**(5), 78–80.
3. Lutz, J. T., Jr., *Plastics Design Proc.* (Sept. 1971) 30–33.
4. Burrell, H., Small, *Interchem. Rev.* (Spring 1955) 3–46.

RECEIVED October 11, 1973.

8

Silane Coupling Agent for Reinforcing Mineral-Filled Nylon

S. E. BERGER, P. J. ORENSKI, and M. W. RANNEY

Union Carbide Corp., Chemicals and Plastics, Tarrytown Technical Center, Tarrytown, N. Y. 10591

Mineral-filled nylon 6 and nylon 6,6 were studied using γ-aminopropyltriethoxysilane to improve the physical properties of the composites and as an aid during processing. The fillers studied included wollastonite (calcium metasilicate) and hydrous kaolin clays, a fibrous inorganic titanate, novaculite silica, and alumina trihydrate. All of these fillers, of widely varying particle size and surface structure, were responsive to the interfacial modification of the aminosilane coupling agent. The improvements in flexural and tensile strength both dry and after water immersion, heat deflection temperature, and, at times, impact strength, are documented and discussed for the mineral filler-polyamide composites.

Polyamide resins are well-established high performance engineering plastics. Characterized by high strength, toughness and good processability, polyamides are challenging metals as structural components in machinery, appliance, and automotive applications. Until recently, they were used either unfilled or with reinforcing fillers such as glass fibers. Although glass fibers enhance strength and impart other benefits, they do not remove what is probably the most serious commercial limitation of glass-filled engineering plastics in general—a high cost/performance ratio.

A sustained effort is now being made to lower this unfavorable ratio by blending high cost engineering plastics, notably nylons, with relatively low cost mineral fillers such as wollastonite, clay, and other similar materials. The resulting composites have mechanical properties that are clearly superior to those of the base resin at an appreciably lower pound-volume cost. Indicative of the intense activity in the field is the recent introduction of mineral-filled nylons as a new class of engineering materials by Monsanto (Vydyne), Dupont (Minlon), and Nypel.

In addition to reducing the cost of the base resin, particulate mineral fillers impart many desirable properties to polyamide composites from physical, mechanical, and processing viewpoints. Reduced sink marks and moisture absorption, greatly increased rigidity, strength and thermal operating range, reduced mold shrinkage, and faster molding cycles are some of the principal benefits.

The high filler content ($\sim$ 50 wt %) required in polyamides to achieve attractive cost/performance ratios results in severe incompatibility at the polyamide/filler interface. This incompatibility seriously reduces adhesion between the two phases and in turn produces poor composite properties, especially after exposure to humid environments.

Silane coupling agents are efficient and practical agents for promoting adhesion between organic resins and mineral fillers, thus overcoming many of the disadvantages associated with high filler loadings. Coupling agents are monomeric silanes of the general structure R^1–$Si(OR)_3$ in which two distinct centers of reactivity exist. The first site, R^1, is a common organofunctional group such as amino, vinyl, epoxy, methacryloxy, mercapto, etc., bonded to the silicon atom by a short alkyl chain. The second reactive site is centered around the silicon atom and consists of hydrolyzable alkoxy groups $Si(OR)_3$.

In use, the alkoxy groups hydrolyze to form silanols (Si–OH) which can react or otherwise condense in the presence of active silica, clay, wollastonite, or metal oxide surfaces. At the other end of the silane molecule, the functional organic groups described above react with suitable groups in the organic matrix resin. To be effective in a given composite, the silane must be reactive to some degree with both the organic and the inorganic components. In practice, the silane can be applied to the filler in a separate pretreatment step, or it can be added directly to the resin and will migrate to the resin/filler interface during processing.

Commencing about 12 years ago, a number of publications (*1, 2, 3*) described the property improvements obtained with silane coupling agents in glass-fiber reinforced thermosetting and thermoplastic polymers. Particulate mineral fillers are used in many composites, and more recent publications (*4, 5, 7, 8*) and a patent (*6*) document property improvements with silane coupling agents analogous to the improvements with glass-fiber reinforced polymer composites.

A previous study (*7*) showed silane coupling agents to be very effective in improving the physical, mechanical, and processing characteristics of highly filled engineering plastics in general and of nylon 6 and nylon 6,6 composites in particular. These earlier results were obtained with commercial resin pellets ground to 50 mesh and then blended with silane-treated filler.

The work described here represents an effort to extend those earlier results and to make them consistent with commercial practice which does not normally use finely ground nylon. Specifically, the effect of the amino-functional silane A-1100 in mineral-filled polyamide resins was studied with as-received pelletized materials using readily available commercial fillers.

Experimental

Typical Compounding and Molding Conditions. PREMIXING. Resin pellets, filler, and silane were thoroughly mixed in a Twin-Shell blender equipped with an intensifier bar. The mixture was dried in an oven for 16 hours at 80°C.

COMPOUNDING. All dried resin/filler/silane mixtures were fused in a 1⅜-inch Egar compounder with zone temperatures controlled as shown below. The extrudate was granulated and dried for 16 hours at 80°C.

Zone 1	475°F
Zone 2	500°F
Zone 3	550°F
Die	550°F

Table I. Fillers Evaluated

Filler	Typical Properties			
	Type	*Shape*	*Particle Size, μ*	*Surface Area, m^2/gram*
Wollastonite P–1 (Interpace)	Calcium silicate	acicular	9 (8:1)[a]	1 (approx.)
Hydrogloss	kaolin clay, untreated	platey	0.3	20–24
Nulok 321 (J. M. Huber)	kaolin clay, aminosilane	platey	0.3	20–24
Optiwhite (Burgess Pigment)	kaolin clay	amorphous	1.4	—
Fybex (Dupont)	potassium titanate	acicular	0.10–0.16 (40:1)[a]	7–10
Novacite L-207A (Malvern Minerals)	novaculite silica	platey	1.3	2
Hydrated alumina C-331 (Aluminum Co. of America)	$Al_2O_3 \cdot 3H_2O$	granular	8–9	<1

[a] Length-to-width ratio.

MOLDING. Test specimen were molded on a 3-oz Van Dorn injection molding machine. Typical molding conditions for 50% composites are shown below. Temperatures were increased 25–100°F for 60% and 70% filled systems.

	Nylon 6	*Nylon 6,6*
Cylinder temp., °F		
rear	515	545
center	530	550
front	550	560
Mold temp, °F.	170	180
Injection pressure, psi	5000	9000
Back pressure, psi	30	50
Screw speed, rpm	96	144
Total cycle time, sec	30	50

Discussion

Using nylon 6 or nylon 6,6 as the resin component, composites were prepared using wollastonite, Fybex, clay (hydrous, calcined, and commercial silane-treated), novaculite silica, and alumina trihydrate (Table I). Clearly, many other mineral fillers could be used to improve the cost and/or performance of nylon for specific applications. However, these seven materials provide a sufficiently broad range of chemical and surface structure as well as physical form to define clearly the contributions of the silane adhesion promoter at many common nylon/mineral filler interfaces.

After extensive screening studies, a single silane, γ-aminopropyltriethoxysilane ($H_2NCH_2CH_2CH_2Si(OEt)_3$), was selected as the most uni-

Table II. Effect of Varying Filler Content for Wollastonite[a]

	Unfilled Control	*50% Wollastonite*	
		Untreated	*Silane*
Flexural strength, psi			
"Initial"	12,600	16,000	18,300
16 hrs in 50°C water	5,400	8,500	11,000
7 days in 50°C water	4,200	4,900	6,600
Standard deviation of "initial"	217	491	238
Flexural modulus, 10^5 psi			
initial	2.9	7.4	6.9
16 hrs in 50°C water	0.9	2.6	2.9
7 days in 50°C water	0.7	1.1	1.7
Tensile strength, psi			
initial	8,900	8,200	10,200
16 hrs in 50°C water	6,200	4,800	4,800
Deflection temp., °C			
264 psi	55	120	120

[a] Grade P-1, Interpace Corp.
[b] Plaskon 8201, Allied Chemical Co.
[c] Pretreated filler with 1 wt % γ-aminopropyltriethoxysilane (A-1100, Union Carbide Corp.) based on filler weight.

formly effective coupling agent for filled nylon systems. Thus, the organic reactivity of the aminofunctional silane with nylon, which may be postulated to occur by transamidation at the processing temperatures used, is constant throughout the study.

The major variable, and the item of most concern technically and practically, is the inorganic reactivity of the silane with various filler surfaces. Furthermore, once having formed the silane–filler interaction product, it is of utmost importance to ascertain the strength of the "bond" under adverse environmental test conditions.

In all studies the polyamide and the aminosilane processing and testing conditions were held constant so that the silane–filler response could be isolated and related to the total physical properties of the various composites. The goals of this study were to determine not only the magnitude of the silane contribution but also to ascertain to what extent the use of the silane would allow the addition of increasing proportions of the low cost fillers leading to a high performance and economically attractive nylon-based composite.

Wollastonite. Wollastonite is a naturally occurring mineral identified chemically as calcium metasilicate. It is mined as acicular-shaped particles which are available in lengths averaging 13–15 times the particle diameter. Our work was based on grade P-1 which has a mean particle diameter of 8.9 microns and a particle length-to-diameter ratio of 8:1.

Filled Nylon 6[b]–Aminofunctional Silane Composites[e]

60% Wollastonite		*70% Wollastonite*		
Untreated	*Silane*[c]	*Untreated*	*Silane*[c]	*Silane*[d]
17,400	21,300	Can't Mold	23,300	can't mold
8,400	14,600		16,300	
5,400	9,400		9,800	
422	275		311	
13.0	12.5	—	14.6	—
3.7	4.9	—	6.8	—
1.1	2.8	—	3.5	—
9,500	11,600	—	11,000	—
5,100	8,900	—	8,000	—
148	165	—	145	—

[d] Integral addition of 1 phf γ-aminopropyltriethoxysilane to resin pellet/filler premix.
[e] Five samples run in each test.

Table III. 50% Wollastonite[a] Filler

	Nylon 6[b]			
	Unfilled	*No Silane*	*Methyl-silane[d]*	*Amino-silane[e]*
Flexural strength, psi				
initial	12,500	16,800	16,900	20,700
16 hrs in 50°C water	5,700	8,500	10,400	11,200
7 days in 50°C water	4,100	5,00	5,500	7,600
Flexural modulus, 10^5 psi				
initial	2.7	7.5	9.1	9.0
16 hrs in 50°C water	0.8	2.7	3.3	2.8
7 days in 50°C water	0.8	1.3	1.4	1.7
Tensile strength, psi				
initial	9,000	8,300	9,600	11,500
16 hrs in 50°C water	6,100	4,000	4,200	6,200
Deflection temp., °C 264 psi	57	113	145	135
Dart impact strength, inches/lb	600	5	7	35

[a] Grade P-1, Interpace Corp.
[b] Plaskon 8201, Allied Chemical Co.
[c] Zytel 101, E. I. duPont de Nemours & Co.
[d] Methyltrimethoxysilane integrally at added 1 wt %, based on filler weight to resin pellet/filler premix.

In keeping with the objective of optimizing cost/performance properties, the effect of varying wollastonite content was studied in nylon 6. To ensure uniformity of data, the silane content was maintained at 1 part (weight) per 100 parts wollastonite. Integral addition of the aminofunctional silane A-1100 was done by premixing nylon pellets, wollastonite, and silane followed by fusion in one pass through a compounding extruder.

The results in Table II indicate the strength improvements obtained with A-1100 relative to untreated wollastonite. The contribution of silane toward improved processing properties is clearly illustrated by the compound containing 70% wollastonite. The results at 70% wollastonite also show a real advantage in using a filler pretreated with silane. In this case it was not possible to mix or mold untreated filler with integrally added silane.

The significant improvements obtained with the aminofunctional silane in a wollastonite-filled nylon 6 composite were also obtained with nylon 6,6 as shown in Table III. The data show the striking property improvements obtained with A-1100 both initially and after water immersion. Addition of wollastonite seriously degrades the impact strength of nylon 6, but the use of aminofunctional silane promotes a significant improvement from 5 inch-lbs for untreated filler to 35-inch-lbs after silane treatment.

Effect of Aminofunctional Silane[e,f]

	Nylon 6,6[c]	
Unfilled	*No Silane*	*Aminosilane*[e]
14,300	13,500	17,100
8,900	9,600	12,400
—	—	—
3.2	10.8	10.7
1.7	6.4	7.0
—	—	—
9,400	7,900	8,000
8,000	6,300	7,600
75	202	215
—	—	—

[e] γ-Aminopropyltriethoxysilane integrally added at 1 wt % based on filler weight to resin pellet/filler premix.
[f] Five samples run in each test.

A non-reactive silane, methyltrimethoxy silane, gives only marginal improvements in filled nylon 6. The magnitude of strength increases noted with this silane are considered typical of those obtained as a result of improved filler dispersion.

The overall advantage of replacing a large part of the high-cost nylon matrix with a low-cost, silane-treated filler is summarized in Table IV.

Wollastonite–Fybex. The effect of replacing part of the wollastonite with Fybex is shown in Table V for 50% mineral-filled nylon 6 and nylon 6,6. Fybex, recently introduced by Dupont, is a fibrous potassium titanate crystal approximately 0.1 micron in diameter and several microns long. The length-to-width ratio is about 40 to 1, and surface area is 7–10 m^2/gram. Fybex-reinforced plastics demonstrate improved properties in electroplating and white pigmentation applications.

Table IV. Property Advantages of Nylon 6 and Nylon 6,6 Filled with Silane-Treated Wollastonite *vs.* Unfilled Nylon

	Improvement, %			
	Nylon 6			*Nylon 6,6*
	50% Filler	*60% Filler*	*70% Filler*	*50% Filler*
Flexural strength				
initial	45	69	85	20
16 hours, 50° C water	104	170	205	39
7 days, 50° C water	57	124	133	
Impact strength	30 inch-lbs			

Table V. 50% Wollastonite + Fybex-Mixed Nylon 6[c]

	Control Unfilled	Untreated	Silane
Flexural strength, psi			
initial	12,500	18,000	21,900
16 hrs in 50°C water	5,700	7,700	13,300
7 days in 50°C water	4,100	5,500	8,900
Flexural modulus, 10^5 psi			
initial	2.7	10.2	10.2
16 hrs in 50°C water	0.8	2.7	3.7
7 days in 50°C water	0.8	1.2	2.0
Tensile strength, psi			
initial	9,000	9,700	11,800
16 hrs in 50°C water	6,100	4,200	8,700
Deflection temp, °C, 264 psi	57	147	125
Dart impact strength, inch-lbs	600	4.5	37

[a] 33% Wollastonite Grade P-1 (Interpace Corp.)–17% Fybex (Dupont).
[b] Integrally added 1 wt % γ-aminopropyltriethoxysilane based on filler weight to resin pellet/filler premix.

Addition of A-1100 silane to a wollastonite–Fybex combination promotes impressive strength improvements in nylon 6 and 6,6. Table VI summarizes the property advantages obtained with silane treatment compared with unfilled nylon.

Kaolin Clay (Hydrated or Calcined). Kaolin clay is a hydrous aluminum silicate mineral with a platey structure. Water-fractionated or hydrous clays are, as a class, more closely controlled for particle size by water fractionation. Kaolin clays contain about 14% water of hydration, most of which can be removed to yield calcined clay which is used when low water absorption and good electrical properties are needed. The use of silane coupling agents with kaolin clay is well known (9), and a number of silane-treated clays are commercially available. Nulok 321 (J. M. Huber Corp.) used in this study is a commercially available aminosilane-treated kaolin clay. Nulok 321 effectively promotes property improvements as a filler in a number of polymers and particularly in elastomers.

Table VI. 50% Wollastonite + Fybex with Silane *vs.* Unfilled Nylon 6 and Nylon 6,6

	Improvement, %	
	Nylon 6	Nylon 6,6
Flexural Strength		
initial	75	56
16 hours, 50° C water	133	78
7 days, 50°C water	117	110

Fillers[a] Effect of Aminofunctional Silane[b,e]

Nylon 6,6[d]		
Control Unfilled	*Untreated*	*Silane*
16,800	19,200	26,200
9,800	10,100	17,500
5,800	7,200	12,200
4.2	14.0	13.7
1.7	5.1	6.0
1.0	1.9	3.3
11,200	11,400	12,700
7,700	6,100	9,400
75	211	211
14	4.5	4.5

[c] Plaskon 8201, Allied Chemical Co.
[d] Zytel 101, Dupont.
[e] Five samples run in each test.

The data in Table VII show significant strength improvements with aminofunctional silanes in hydrous or calcined clay composites of nylon 6. The readily available silane-treated Nulok 321 clay composite shows clear advantages in strength over the no-silane control and, most noticeably, over the unfilled nylon 6 matrix. The last point has important economic consequences. Using current market-place price data, this result indicates that one can replace a 3.1¢/in^3 resin with a 2.4¢/in^3 composite in an application where impact strength is not a determining factor and achieve not only a cost advantage but also a definite performance benefit, especially under unfavorable high temperature and humidity.

Novaculite Silica. Novaculite silica is a highly pure, finely granular form of silica rock. Because of its platey structure, it is free of jagged edges and reportedly therefore less abrasive than common types of silica. Novakups, a silane-treated novaculite silica, has recently been made commercially available by Malvern Minerals Co. Table VIII shows the superior strength levels which can be obtained with A-1100 treated novaculite silica in nylon 6 or nylon 6,6 composites. As expected from both theory and practice, the effect of coupling agent on composite strength is most noticeable after exposure to high humidity. Gains of over 100% in wet flexural and tensile strength of silane-containing nylon 6,6 composites *vs.* an untreated control are particularly noteworthy. As established in a previous study (8) silane-treated filler increases the heat deflection temperature of nylon 6,6 composites relative to a control and

Table VII. Kaolin Clay-Filled–Nylon 6[a, f] Effect of Aminofunctional Silane

	Unfilled Control	50% Calcined Clay		50% Hydrous Clay	
		Un-treated[b]	*Silane[c]*	*Un-treated[d]*	*Silane[e]*
Flexural strength, psi					
initial	12,500	17,300	21,900	14,900	19,300
16 hrs in 50°C water	5,700	10,000	14,500	10,200	14,100
7 days in 50°C water	4,100	6,600	9,600	—	—
Flexural modulus, 10^5 psi					
initial	2.7	8.5	9.1	7.9	7.2
16 hrs in 50°C water	0.8	3.5	3.7	4.8	5.0
7 days in 50°C water	0.8	1.8	2.0	—	—
Tensile strength, psi					
initial	9,000	10,200	11,300	8,300	10,600
16 hrs in 50°C water	6,100	6,200	8,900	5,900	7,200
Deflection temp °C, 264 psi	57	84	128	153	151
Dart Impact strength, inch-lb	600	5.5	17.5	4.5	4.5

[a] Plaskon 8201, Allied Chemical Co.
[b] Optiwhite, Burgess Pigment Co.
[c] Pretreated with 1.0 wt % γ-aminopropyltriethoxysilane based on filler weight.
[d] Hydrogloss, J. M. Huber Corp.
[e] Nulok 321, J. M. Huber Corp.
[f] Five samples run in each test.

Table VIII. 50% Novaculite Silica[a]

	Nylon 6[c]		
	Control Unfilled	*Untreated*	*Silane*
Flexural strength, psi			
initial	12,500	15,500	17,700
16 hrs in 50°C water	5,700	7,400	10,100
7 days in 50°C water	4,100	5,300	7,200
Flexural modulus, 10^5 psi			
initial	2.7	8.0	7.1
16 hrs in 50°C water	0.8	2.7	2.3
7 days in 50°C water	0.8	1.0	1.3
Tensile strength, psi			
initial	9,000	8,300	9,300
16 hrs in 50°C water	6,100	4,600	6,700
Deflection temp, °C, 264 psi	57	128	114
Dart impact strength, inch-lb	600	4.5	17

[a] Novacite L-207A, Malvern Minerals Co.
[b] 1.0 wt % γ-aminopropyltriethoxysilane based on filler weight integrally added to resin pellet/filler premix.

has a significant positive effect on the dart impact strength of filled nylon 6 material (Table VIII). Work is continuing with many other mineral fillers, chopped glass, roving, and some new silane-containing polymeric film forming compositions to optimize the strength relationships in the filler–polymer interfacial region and thereby obtain high performance composites at the lowest possible cost. Studies have either been conducted or are underway with glass beads, beach sand, other commercial silicas and clays, and inorganic flame-retardant additives.

Alumina Trihydrate. Silane coupling agents are now used in many rubber and thermosetting resin products which contain alumina trihydrate. In such applications, the trihydrate is used in amounts up to several hundred parts per 100 parts base polymer, depending on need for improved arc tracking and/or flame retardance. Although alumina trihydrate is not typically used in nylon, some preliminary data have been obtained to determine the silane response of these relatively weak hydrogen bonded particles. Table IX shows that definite improvements in physical properties are obtained by using A-1100 aminosilane in a 50% alumina trihydrate nylon 6 composite. The strength retention after water conditioning is particularly surprising in view of the structure of the trihydrate.

Variability of Test Results. All mechanical property data in this study were obtained with standard ASTM test methods, and the average of five observations is reported. Standard deviation of "initial" flexural

Filler Effect of Aminofunctional Silane[b, e]

	Nylon 6,6[d]	
Control Unfilled	*Untreated*	*Silane*
16,800	10,500	16,500
9,800	6,000	13,400
5,800	4,800	9,700
4.2	9.2	9.4
1.7	3.4	4.7
1.0	1.4	2.2
11,200	4,600	10,100
7,700	3,300	7,400
75	114	177
14	4.5	4.5

[c] Plaskon 8201, Allied Chemical Co.
[d] Zytel 101, Dupont.
[e] Five samples run in each test.

Table IX. Nylon 6–Alumina Trihydrate Filled Effect of A-1100 Treatment

	Control, Unfilled	*50% Hydral C-331*	
		Untreated	*1 phf A-1100*
Flexural strength, psi			
initial	12,000	11,000	13,900
16 hrs, 50°C H_2O	5,700	7,400	9,000
Flexural modulus, 10^5 psi			
initial	2.7	6.1	5.2
16 hrs, 50°C H_2O	0.8	2.0	2.2

strength values are shown in Table II. These values are typical of all composites considered here. The significantly lower standard deviation of composites containing aminosilane treated filler compared with untreated filler suggests more uniform properties and better filler dispersion.

Conclusion

Nylon 6 or nylon 6,6 can be highly filled (up to 70 wt %) when an aminosilane coupling agent is used to modify the mineral filler–resin interface. The flexural and tensile strengths of the composites, both dry and wet, are significantly increased using a variety of mineral fillers. In some formulations, the heat deflection temperature is also improved.

Overall, the composites evaluated provided a lower cost product with better physical properties and good processing characteristics compared with the base resin. Only the impact strength is adversely affected by adding an untreated filler. Even here, impact strength is somewhat improved in those composites containing aminosilane treated filler. While other properties must be considered for specific end uses, this study provides the basic technology for the continued evaluation of various mineral filled nylons in many applications.

Literature Cited

1. Sterman, S., Bradley, H. B., "A New Interpretation of the Glass-Coupling Agent Surface Through Use of Electron Microscopy," *SPE Trans.* (Oct. 1961).
2. Sterman, S., Marsden, J. G., "Silane Coupling Agents as Integral Blends in Resin Filler Systems," *18th Ann. SPI Preprints* (Feb. 1963).
3. Plueddmann, E. P., "Evaluation of New Silane Coupling Agents for Glass Fiber Reinforced Plastics," *17th Ann. SPI Preprints* (Feb. 1962).
4. Ziemianski, L. P., "A Survey of the Effect of Silane Coupling Agents in Various Non-Glass Filled Thermosetting Resin Systems," *22nd Ann. SPI Preprints* (March 1966).
5. Sterman, S., Marsden, J. G., "The Effect of Silane Coupling Agents in Improving the Properties of Filled or Reinforced Thermoplastics," Part I, *21st Ann. SPI Preprints* (March 1965); Part II, *21st Ann. SPI Preprints* (Feb. 1966).

6. U.S. Patent **3,419,517** (Dec. 31, 1968).
7. Ranney, M. W., Berger, S. E., Marsden, J. G., "Silane Coupling Agents in Particulate Mineral-Filled Composites," *27th Ann. SPI Preprints* (Feb. 1972).
8. Orenski, P. J., Berger, S. E., Ranney, M. W., "Silane Coupling Agents—Performance in Engineering Plastics," *28th Ann. SPI Preprints* (Feb. 1973).
9. Grillo, T. A., "Silane Modified Kaolin Pigments," *Rubber Age* (Aug. 1971).

RECEIVED October 11, 1973.

9

Catalytic Effects in Bonding Thermosetting Resins to Silane-Treated Fillers

EDWIN P. PLUEDDEMANN

Dow Corning Corp., Midland, Mich. 48640

Mineral surfaces generally inhibit the cure of thermosetting resins. From measurements of cure exotherms of resin castings filled with silane-treated minerals, it is concluded that silane coupling agents increase the cure to "tighten up" the resin structure at the interface. Best strength and water resistance of organic–mineral composites, therefore, require a "restrained layer" rather than a "deformable layer" at the interface. A silane-treated zircon is recommended as an outstanding filler for polyester resins.

Optimum mechanical performance of a mineral-reinforced organic polymer composite imposes contradictory requirements on the interface between the polymer and the mineral:

(1) Optimum stress transfer between a high modulus filler and a lower modulus resin requires an interphase region of intermediate modulus.

(2) Composite toughness and ability to withstand differential thermal shrinkage between polymer and filler require a flexible boundary region or a controlled fiber pull-out to relieve localized stresses.

Restrained Layer Theory

Kumins and Roteman (*1*) suggested in 1963 that the boundary region, of which the coupling agent is a part, between a high modulus reinforcement and a lower modulus resin can transfer stresses most uniformly if it has a modulus intermediate between that of the resin and the reinforcement. In agreement with this, it has generally been observed that polymer molecules adsorbed on a rigid filler particle are more closely packed than in the bulk, and the degree of ordering decreases with distance from the reinforcement. Kwei proposed that the sphere of influence of the filler particles can be as great as 1500 A, and there will be a regular gradation of mechanical properties over this distance (*2*). Schonhorn *et al.* have demonstrated that adhesion of a rigid epoxy resin to polyethylene can be improved markedly if the rigidity of the polyethylene surface is increased by irradiation in a glow discharge (*3*) or by crystallizing the polymer in contact with gold (*4*). The restrained layer theory suggests that silane coupling agents function by "tightening up" the polymer structure at the interface while simultaneously providing silanol groups for bonding to the mineral surface (*5*). It is difficult to reconcile this concept with the need for stress relaxation at an interface because of differential thermal shrinkage between polymer and filler.

Deformable Layer Theory

Hooper proposed in 1956 that the silane finish contributed mechanical relaxation through a deformable layer of silicone resin (*6*). It was soon apparent, however, that the layer of silane in a typical glass finish was too thin to provide stress relaxation through mechanical flexibility.

A preferential adsorption theory proposed by Erickson *et al.* (*7*) is a modification of the deformable layer theory. This new theory was based on the assumption that different finishes on glass fibers have, to different degrees, the power to deactivate, destroy, or adsorb out of the uncured liquid resin mixture certain constituents necessary to complete resin curing. This would lead to an upset in the optimum local material balance by what is termed "preferential adsorption." It was assumed that this effect was important only near the surface since the separation process would depend on diffusion rates, which would be low in viscous resins. Carried to its logical conclusion, the above theory implies that a finish would lead to an interface resin layer of variable thickness and flexibility. This postulated flexible layer did not depend on finish thickness and could have a thickness much greater than 100 A. In addition, such a layer would need ductility and strength to provide relaxation and effective transfer to stress between the fibers in load-bearing situations.

Mineral fillers have a catalytic effect (positive or negative) on resin cure, and the effect is specific for each resin and each curing system. Thus, polyester resins cured with a benzoyl peroxide initiator are less sensitive to mineral surfaces than the same resin cured with a cobalt-promoted ketone peroxide initiator (*8*). Glass fibers treated with a chrome finish retard gelation of a polyester more than glass treated with a silane finish (*9*). It was recently reported that clean glass caused a marked reduction in polyester exotherm (*10*). Silane and chrome coupling agents on the glass restored a major portion of the exotherm, better coupling agents (as indicated by mechanical performance of composites) giving the higher exotherms. These data support the "restrained" layer theory for optimum composite performance.

Reversible Hydrolyzable Bond Theory

A reversible hydrolyzable bond mechanism proposed by Plueddemann (*10*) combines the features of a chemical bond theory with the rigid interface of the restrained layer theory while allowing the stress-relaxation properties of a deformable layer theory. It is proposed that reversible breaking and remaking of stressed bonds between coupling agent and glass in the presence of water allows relaxation of stresses without loss of adhesion. Water resistance is possible under conditions of hydrolytic equilibrium if the resin at the interface is rigid but not if it is rubbery. Since stress relaxation by reversible hydrolysis at the interface is limited to molecular dimensions, this mechanism is most effective in resin composites with fine fibers or particulate fillers in which strains in the resin between particles are limited to small distances approaching molecular dimensions.

Experimental and Results

Heat-cleaned style 7781 E-glass cloth and common particulate minerals were compared as fillers in three resin systems. Silanes of Table I were applied to glass cloth and to glass microbeads from aqueous solution. Experimental silane-treated clays were obtained from J. M. Huber Co. Silane-treated silicas (Novacite 1250) were supplied by Malvern Minerals Co. Resin systems included:

Polyester	*A*	*B*
Paraplex P-43	100	100
Styrene	10	10
Benzoyl peroxide	1	—
MEK-peroxide "60"		1.67
1% Co in styrene		1.0

Exotherm in water bath at 90°C

Epoxy	*A*	*B*
DER-330	100	100
Curing agent "Z"	18	—
Aminoethylpiperazine	—	12

Exotherm in oil bath at 115°C

Urethane Prototype	
Dipropylene glycol	13.4 grams
Toluene diisocyanate	17.4 grams

Mix in open vial at 25°C, no bath.

Table I. Organofunctional Silane Coupling Agents[a]

Organofunctional Group	*Structure of Organofunctional Group*
Amine	$—CH_2CH_2CH_2NH_2$
Diamine	$—CH_2CH_2CH_2NHCH_2CH_2NH_2$
Polyamino	Dow polyethyleneimine
Epoxy	$—(CH_2)_3OCH_2\overset{O}{CH—CH_2}$ (epoxide ring)
Mercaptan	$—CH_2CH_2CH_2SH$
Cationic styryl	$—CH_2CH_2CH_2NH—CH_2CH_2NH$ V.B.·HCl[b]
Methacrylate ester	$—CH_2CH_2CH_2$M.A.[c]
Methacrylate quaternary ester	$—(CH_2)_3\overset{+}{N}(CH_3)_2—CH_2CH_2$M.A.[c] Cl^-
Saturated quaternary alcohol	$—(CH_2)_3\overset{+}{N}(CH_3)_2—CH_2CH_2OH$ Cl^-

[a] Silanes supplied by Dow Corning Corp.
[b] V.B. = vinylbenzyl.
[c] M.A. = methacrylate.

The catalytic effects of mineral surfaces were first examined by measuring the maximum exotherms of catalyzed resin systems containing 5 grams of filler in 30 grams of resin. This low filler level allowed observation of surface catalytic effects without serious distortion by the masses of inert minerals of different heat capacities.

Table II. Exotherm Changes in the Presence of Fillers

5 grams Filler in 30 grams Resin

Resin System	*Polyester–Bz_2O_2*	*Epoxy "Z"*	*Glycol–TDI*
Bath Temp., °C	*90°*	*115°*	*25°*
Control Max. Temp.	*225°*	*240°*	*155°*
Filler	Δ *T, °C*	Δ *T °C*	Δ *T, °C*
Barium sulfate, powder	−3	—	−24
Calcium carbonate[a]	−6	−9	−38
Silica, Novacite 1250	−8	−22	−46
Calcium silicate[b]	−8	−24	−37
Talc WCD 2801	−8	−30	−20
Alumina, Lima LPA-12	−9	−8	+40
Zircon, milled TAM	−11	−34	−5
E-glass cloth	−15	−13	+2
A-glass microspheres	−12	−5	+20
Mica −325 mesh	−15	−21	−33
Graphite[d]	−43	−28	−49

[a] Av. 5-μ Snowflake Whiting, Thompson Weinman & Co.
[b] Av. 4-μ wollastonite P-1, Interpace Corp.
[c] 10–50-μ Glasshot MS-XL, Cataphote Corp.
[d] Dixon-0525.

Maximum exotherm temperatures reached were, in each case, compared with the exotherm of an unfilled resin reaction (Table II). Except for a few minerals in a urethane system, all fillers lowered the maximum exotherm. The degree of change in exotherm differed with each filler

Table III. Polyester Exotherms in the Presence of Fillers

Filler in 20 grams Resin

	Maximum Exotherm in 90°C Bath			
	10 grams Filler		*20 grams Filler*	
Filler	Δ *Time, sec*	Δ T, *°C*	Δ *Time, sec*	Δ T, *°C*
Zircon	−35	−7	−24	−26
Barium sulfate	−59	−12	−32	−31
Calcium carbonate	−15	−27	−7	−45
Wollastonite	−11	−17	+28	−47
Mica	+10	−20	—	—
Silica	+6	−25	+32	−51
Glass microspheres	+4	−19	+38	−36
Talc	+24	−28	+26	−62
Alumina	+50	−13	+49	−45
Clay	+49	−30	+53	−57

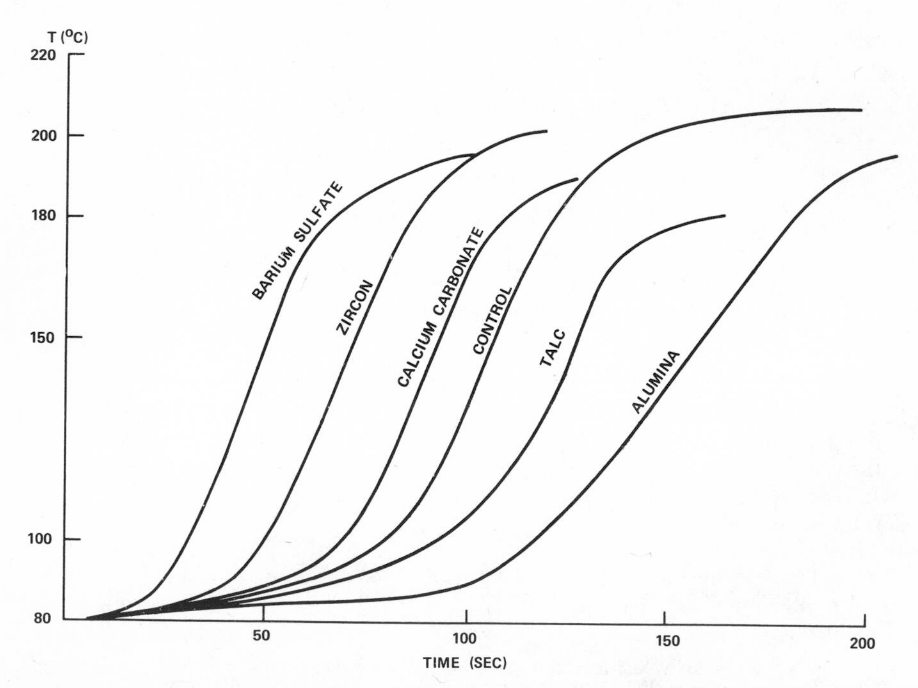

Figure 1. Polyester exotherms; 10 grams of filler in 20 grams of resin in 90°C bath.

in each resin system, indicating a specific catalytic effect superimposed on an inert dilution effect. For a given filler–resin system the exotherm depression was proportional to surface area of the filler.

Minerals were studied more completely in the polyester resin at 33% and 50% filler loadings (Table III, Figure 1). Change in cure time (from 80° to 150°C) was not always accompanied by a comparable change in exotherm. Zircon, barium sulfate, and calcium carbonate promoted a much faster cure than an unfilled resin. Alumina retarded the cure but still allowed a creditable exotherm. All fillers lowered the maximum exotherm. With some (clay, talc, and silica) this inhibition of cure is severe enough to limit their usefulness in highly filled systems.

The effects of silane treatments on fillers were studied separately for each resin system. Only a limited amount of glass cloth could be dispersed uniformly in a resin casting and was studied with 5 grams of glass in 30 grams of resin. As reported earlier, silane coupling agents on glass generally restored some of the lost exotherms in polyester and epoxy resins (Table IV). Correlation with properties of glass-reinforced resin composites suggest that silane treatments that give the highest exotherms also provide composites with best mechanical properties and chemical resistance.

Table IV. Resin Exotherms with Silane-Treated Glass

5 grams E-Glass Cloth in 30 grams Resin

Organofunctional Silane 0.2% on Glass	*Polyester–Bz_2O_2* Δ T, °C	*Epoxy–"Z"* Δ T, °C
Untreated	−22	−17
Epoxy	−20	−8
Diamine	−15	−1
Mercaptan	−11	−8
Cationic styryl	−10	0
Phenyl	−7	−15
Methacrylate ester	−3	—

Similar studies were made with silane-treated A-glass microbeads at higher filler loadings in polyester resin cured with benzoyl peroxide or with methyl ethyl ketone (MEK) peroxide and cobalt promoter (Table V). Although silane treatments on glass allowed a faster rate of cure and a higher exotherm, the effect was not very pronounced. Reactive silanes containing a mercaptan, a methacrylate ester, or a cationic styryl group looked best in the polyester cured with benzoyl peroxide. Functional silanes containing quaternary ammonium functions looked best in the polyester cured with MEK peroxide and cobalt promoter. This is not surprising since quaternary ammonium compounds are known activators for cobalt-promoted peroxide initiation.

Table V. Polyester Exotherms with Silane-Treated Glass

20 grams A-Glass in 20 grams Resin

Organofunctional Silane	*Benzoyl Peroxide*		*MEK Peroxide + Cobalt*	
	Δ *Time, sec*	Δ T, °*C*	Δ *Time, sec*	Δ T, °*C*
(Control, No Glass)	(109)	(219)	(95)	(220)
Untreated	+40	−51	+60	−54
Methacrylate ester	+5	−35	+57	−48
Methacrylate quaternary salt	+35	−42	+57	−42
Cationic styryl	+38	−40	+52	−47
Mercaptan	+22	−47	—	—
Saturated quaternary salt	+24	−37	+40	−37

Silane-treated silicas were compared at relatively low filler loading in epoxy resins cured with aromatic amines (agent "Z") or aliphatic amines (aminoethylpiperazine) (Table VI). All of the coupling agents were very effective in restoring the epoxy exotherm with aromatic amine cure. The cationic styryl silane was, by far, the most effective with aliphatic amine cure.

Table VI. Epoxy Exotherms with Silane-Treated Silica

5 grams Silica in 30 grams Resin

Organofunctional[a] *Silane (0.5% on Silica)*	*Curing Agent "Z"* Δ T, °*C*	*Curing Agent AEP*[b] Δ T, °*C*
Untreated	−22	−25
Amine	−6	−30
Diamine	−9	−13
Polyamine	−2	−31
Cationic styryl	−6	+2
Epoxy	−6	−39

[a] Novacite 1250 treated with 0.5% silane; Malvern Minerals Co.
[b] AEP; 12 parts aminoethylpiperazine in 100 parts DER-330.

Silane-treated clays were compared both in an epoxy–"Z" resin and in the urethane prototype (Table VII). The mercaptofunctional silane was best with both systems. A cationic styryl silane should also be tested on clay fillers since it is a promising treatment on glass and on silica.

Discussion

Exotherms of catalyzed resins cured in the presence of mineral fillers demonstrate that many mineral surfaces inhibit resin cure. Loss in exotherms attributed to mineral surfaces may be partially restored by treating the surface with a suitable silane. Mechanical properties of composites

Table VII. Exotherm Depression with Silane-Treated Clay[a]

5 grams Clay in 30 grams Resin

Organofunctional Silane 1.0% on Clay	*Epoxy "Z"* Δ T, °C	*Urethane* Δ T, °C
Untreated	−41	−34
Diamine	−38	−18
Mercaptan	−15	−12
Epoxy	−32	−27

[a] Kaolin clay from J. M. Huber Corp.

are better with silane-treated minerals that allow resins to cure with higher exotherms. A restrained layer rather than a deformable layer must, therefore, be the better morphology at the filler–resin interface for optimum mechanical properties and water resistance of composites.

Resin morphology resulting from complete cure at the interface is only one factor in determining composite properties. A phenylsilane on glass, for example allows polyesters to cure with full exotherm, but it is not an effective coupling agent since it does not react with the resin. The second factor in determining composite properties is the co-reaction of the silane with the resin so that the resin presents a silanol surface to the mineral. The silanol-modified resin bonds to the mineral through a hydrolyzable equilibrium that imparts ductility while maintaining bond integrity in the presence of water.

Silane treatment of certain fillers may not restore sufficient reactivity at the resin interface to allow high filler loadings of the resin in practical systems. In this case additional means must be used to restore resin cure. This is probably more important with free-radical induced cure than with condensation cure since extra post-cure is relatively ineffective after the free radical initiator is consumed.

A practical application of these observations may be made by comparing three fillers in polyester castings. Silica responds well to silane coupling agents, but even with a silane finish it does not allow an adequate exotherm in polyester resin for adequate room temperature cure. Calcium carbonate allows a good polyester exotherm, but it does not respond to silane treatment. Zircon allows a very good polyester cure and responds well to silane treatment. Fillers treated with 0.2% cationic styryl-functional silane are compared with untreated fillers at 60% loadings in polyester castings (Table VIII). Treated zircon should be an ideal filler for polyesters either alone, or mixed with silane-treated fillers that normally give poor cure (*e.g.*, silica).

The cationic styryl-functional silane appears to be a preferred silane on minerals in epoxy resins or in benzoyl peroxide-initiated polyesters. The quaternary ammonium functional methacrylate ester is recommended

Table VIII. Polyester Castings with 60% Mineral Filler

		Flexural Strength, psi (MN/m^2)			
		Untreated		0.2% Silane Treated	
Filler	Cure	Dry	Wet	Dry	Wet
Silica	poor	9,700 (67)	6,700 (46)	19,600 (135)	17,700 (122)
Calcium carbonate	good	11,100 (72)	10,500 (72)	11,000 (76)	10,000 (69)
Zircon	good	11,300 (78)	8,800 (61)	20,000 (138)	15,600 (108)

on fillers in polyester resins cured with ketone peroxides and cobalt promoter. The mercaptofunctional silane is most effective on kaolin in urethanes and epoxies. Improved mechanical and electrical properties and chemical resistance imparted to composites by silane treatment of fillers have been described adequately (*11*). Exotherm measurements on resin cure with treated fillers are a simple method of selecting the optimum treatment for a particular system.

Literature Cited

1. Kumins, C. A., Roteman, J., *J. Polym. Sci.* (1963) **1** (1A), 527.
2. Kwei, T. K., *J. Polym. Sci.* (1965) **3** (3A), 3229.
3. Schonhorm, H., Hansen, R. N., *J. Appl. Polym. Sci.* (1967) **11**, 1461.
4. Schonhorn, H., Ryan, F. W., *J. Polym. Sci.* (1968) **6** (A2), 231.
5. Hartlein, R. C., *Ind. Eng. Chem., Prod. Res. Develop.* (1971) **10** (1), 92.
6. Hooper, R. C., *Proc. SPI 11th Ann. Tech. Conf.* (1956), Sec. 8-B.
7. Erickson, P. W., *Proc. SPI 27th Ann. Tech. Conf.* (1970), Sec. 13-A.
8. Strolenberg, K. *et al.* in "Glasfaserverstaerkte Kunstoffe," P. H. Selden, Ed., p. 139, Springer-Verlag, Berlin, 1967.
9. *Ibid.*, p. 137.
10. Plueddemann, E. P., Stark, G. L., *Proc. SPI 28th Ann. Tech. Conf.* (1973), Sec. 21-D; *cf.* "Interface Phenomena in Polymer Matrix Composites," E. P. Plueddemann, Ed., Chap. VI, Academic, New York, in press.
11. Ranney, M. W., Berger, S. E., Marsden, J. G., *Proc. SPI 27th Ann. Tech. Conf.* (1972), Sec. 21-D.

RECEIVED October 11, 1973.

10

The Rheology of Concentrated Suspensions of Fibers

I. Review of the Literature

RICHARD O. MASCHMEYER and CHRISTOPHER T. HILL

Materials Research Laboratory, Washington University, St. Louis, Mo. 63130

The sparse literature on the rheological properties of concentrated suspensions is reviewed. Various authors have obtained widely different flow curves and have made conflicting observations regarding yield stresses, fiber breakage, extrusion force fluctuations, and elastic effects. Most of the differences can probably be attributed to failure to control or measure fiber length distribution in suspension. Recent work in our laboratory is summarized, and suggestions are made for both experimental and theoretical research needed.

A large portion of short-fiber reinforced thermoset and thermoplastic materials are processed by flow molding techniques including transfer-, injection-, and compression molding. Each processing technique involves the flow in complex geometries of suspensions of short fibers (usually glass) and/or rigid particles in fluids which are polymer melts or liquid prepolymers. These suspensions are usually highly concentrated (10–50% solids by volume), and they often contain trace amounts of an immiscible second phase such as water, a lubricant, or a wetting agent.

The rational design of molding equipment including plasticating devices, molding machine runners and gates, and molds requires a knowledge of the flow properties of the material to be molded at the processing conditions of temperature, pressure, and shear rate. These properties depend on material variables such as matrix rheology; fiber length, stiffness, and strength; volume fraction of fibers and/or particulate fillers; and the nature and amount of wetting agents, lubricants, or other additives. Since we cannot evaluate the flow properties for every material of interest, we would prefer sufficient empirical knowledge and theoretical support to be able to predict the flow behavior of commercial molding compounds with some certainty.

Most of the data available in this area are confounded by problems with resin and/or fiber degradation or lack of control of one or more important variables. In this laboratory we are developing the requisite empirical background concerning the rheology of concentrated fiber suspensions necessary to construct a well-founded model of reinforced plastics processing.

Significance of Rheological Measurements in Concentrated Suspensions

Viscosity is a true material parameter only for homogeneous materials, and short-fiber-reinforced plastics can seldom be treated as homogeneous. Thus, the measured viscosity of a suspension may depend upon both the flow geometry and the geometry of the suspended material. A particular suspension may have different properties in similar geometries of different size—*e.g.*, the viscosity of a suspension in capillary flow may depend upon the diameter of the capillary.

The dependence of viscosity on geometry is the result of both the orientation of the fibers during flow and the interaction of particles with the wall of the measuring instrument. Wall interaction, which has been

Table I. Previous Measurements on

Research Group	*Viscometer*	*Viscometer Dimension (inch) (dia/gap)*	*Matrix Material*	*Fiber Type and Length, inch*
Stankoi *et al.* (*24*)	capillary	0.35–0.47	polyester and kaolin clay	glass, 0.8
Thomas & Hagen (*25*)	capillary	0.06	polypropylene	glass, 0.18–0.44
Newman & Trementozzi (*26*)	capillary	0.5	styrene–acrylonitrile (copolymer)	wollastonite, 10^{-3}
Carter & Goddard (*27*)	cone & plate	0.16	polybutene oil	glass, 0.008–0.03
Mills (*28*)	capillary	?	polyethylene	glass, 0.25
Bell (*29*)	capillary	0.12–0.25	B-staged epoxy	glass, 0.125
Ziegel (*30*)	couette	?	various polymers, liquid at RT	glass, 0.24
Takano (*31–32*)	capillary	1 to 1/4 rectangular	B-staged epoxy	glass, 0.125–0.5
Karnis *et al.* (*33*)	couette & capillary	0.076–0.40	castor oil	nylon, 0.049

thoroughly studied for spheres and dilute suspensions, is described in terms of (a) mechanical interaction involving physical contact of the particle with the wall (*1, 2*); (b) hydrodynamic interaction, in which the presence of the wall alters the velocity profile around the particle (*3, 4*); and (c) radial migration in which particles in tube flow migrate either toward or away from the flow axis (*5, 6, 7, 8, 9*).

Although the magnitude of these effects in dilute fiber suspensions does not seem large (*10*), their importance in concentrated suspensions of fibers is unknown. Thus, although summarizing short-fiber reinforced plastic flow data in terms of viscosity is useful for interpretation in terms of past experience, the data may be valid only for the geometry in which it was measured.

Concentrated Suspension Viscosity Data

The flow of concentrated suspensions of spherical particles has been studied for many years (*11, 12, 13*). While classical studies were concerned primarily with the effect of volume fraction of solids on suspension viscosity, more recent work has considered effects of particle size and particle size distribution (*8, 14*), wetting and second fluid phases (*15, 16, 17*), and viscoelasticity of the fluid phase (*18, 19, 20, 21, 22, 23*).

Concentrated Suspensions of Fibers

Aspect Ratio	*Volume Fraction, %*	*Shear Rate, sec^{-1}*	*Force Fluctuation*	*Yield Stress*	*Fiber Breakage*
?	15	5–150	yes	yes	no
?	0–40	0.38–386	yes	no	perhaps
10	0–50	3–3000	?	?	?
52–228	0.5–2.2	0.167–167	?	no	no
400	20	30 & 900	?	yes	yes
200	45–63	8–100	yes	?	?
500	2	0.5–1.2	?	yes	
200–800	0–60	?	yes	yes	yes
8	8	?	no force measurements		?

A synopsis of experimental literature on the rheology of concentrated fiber suspensions done outside our laboratory is shown in Table I. The columns are mostly self-explanatory; force fluctuation refers to the observation by some that the force required to extrude fiber suspensions through capillaries fluctuates with time. Insufficient data were available to compare the flow curves from all researchers on the same basis, so the qualitative comparison of Figure 1 was constructed. Only the relative shapes and shear-rate ranges are significant.

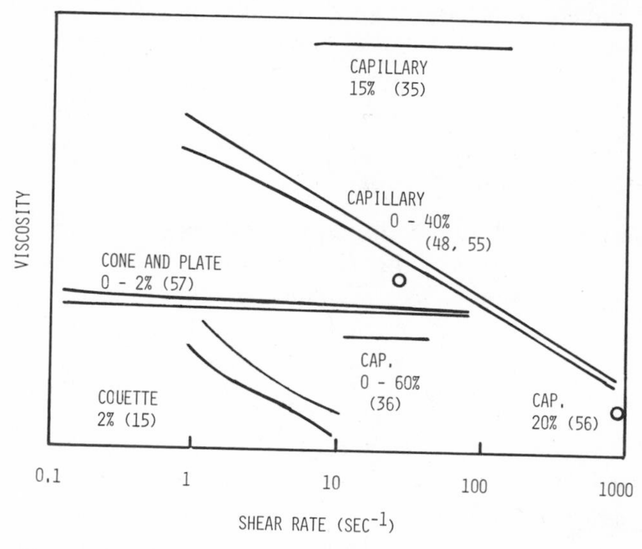

Figure 1. Qualitative comparison of viscosity data for suspensions of fibers

None of these works provide a firm basis for understanding melt-state rheology of reinforced plastics. The thrust of the works of Bell (*29*), Takano (*31, 32*) and Karnis *et al.* (*33*) was to measure the effect of various parameters on fiber orientation in flow, and no quantitative viscosity data were published. The data of Mills (*28*) are fragmentary since only two capillary viscosity points for fiber suspensions were published. Carter and Goddard (*27*) and Ziegel (*30*) measured suspensions with fiber concentrations well below those of commercial interest. Newman and Trementozzi (*26*) were primarily interested in the effects of fillers on die swell. They published one specific viscosity-*vs.*-fiber concentration curve although the fibers involved were quite small and had low aspect ratios.

Thomas and Hagen (*25*) did a more thorough study of short-fiber reinforced thermoplastic rheology with variables in the region of commercial interest for suspensions of glass fibers in polypropylene. Their flow curves fit a power law model, and no yield stresses were observed.

Microscopic inspection of the extrudate showed no fiber migration and possible fiber breakage only at the highest shear rates. Their quantitative results, however, are obscured by resin degradation incurred in mixing.

Stankoi *et al.* (*24*) measured viscosities of suspensions of glass fibers and kaolin clay in a polyester prepolymer. Their flow curves, which displayed yield stresses, were consistent with a Bingham plastic model. They observed no fiber migration or breakage. Since they did not report matrix resin rheological data, it is difficult to generalize their quantitative data to other systems.

Elastic Effects and Normal Stresses in Concentrated Suspensions

Contradictory observations have been made concerning elastic and normal-stress-driven phenomena in fiber suspensions. Newman and Trementozzi (*26*) found that the addition of wollastonite filler to a viscoelastic resin greatly reduced the capillary die swell. Carter and Goddard (*27*) detected no phase lag in dynamic oscillatory testing of a suspension of short fibers in a cone and plate instrument, but they measured large primary normal stress differences. As discussed later, Roberts (*34, 35*) obeserved massive Weissenberg rod climbing and large capillary entrance corrections for a suspension (glass fibers in an inelastic oil) which displayed no die swell and no elastic recovery when rapidly deformed.

Recent Study of Fiber Suspension Rheology

The review above indicates that there is little agreement among researchers about the flow behavior of concentrated fiber suspensions, even on a qualitative level. Their results do suggest, however, the following points:

(a) Concentrated fiber suspensions are highly non-Newtonian, and they may have high yield stresses

(b) Force *vs.* displacement curves in capillary rheometers show fluctuations, probably from log-jamming at the capillary entry

(c) Flow properties may be sensitive to fiber orientation and, hence, to viscometer geometry

(d) Fiber breakage during flow can be severe and cause large changes in suspension viscosity

(e) Brodnyan's theory (*36*) of fiber suspension rheology predicts viscosities which are orders of magnitude too large for concentrated suspensions of high aspect ratio fibers

(f) Observations of elastic effects are contradictory.

In view of the difficulties in working directly with short-fiber reinforced plastics, we have studied the rheology of a model system of glass fibers in silicone oil (*34, 35, 37, 38, 39*). Viscosity measurements were done with the large bore capillary rheometer shown in Figure 2. This device, which is designed for room temperature operation, allows measurement of steady flow viscosity and capillary entrance corrections over the shear rate range 0.055–5500 sec^{-1}. Capillaries are available with length-to-diameter ratios from 1 to 32 and diameters of ⅛ and ¼ inch.

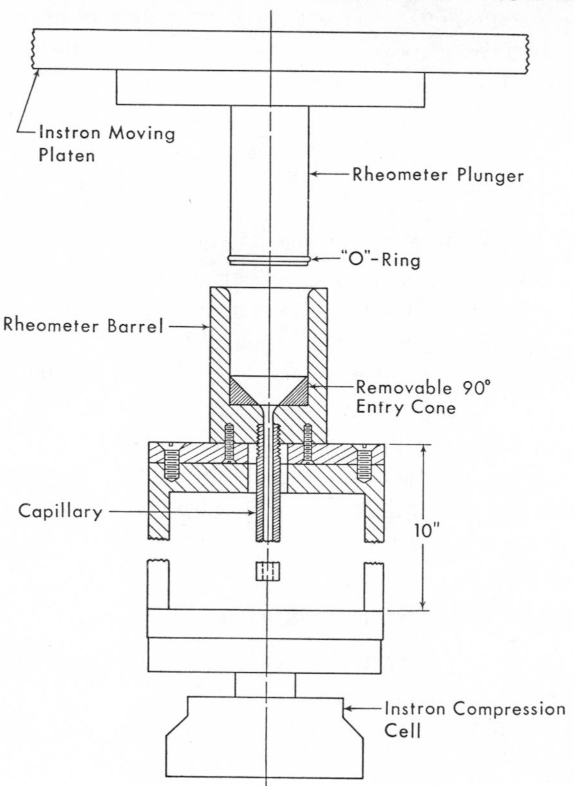

Figure 2. Capillary viscometer

Typical non-Newtonian viscosity curves for 15 and 30 v/o suspensions of glass fibers in 600-poise silicone oil are shown in Figure 3 as obtained in our rheometer (*39*). These data were corrected for capillary entrance losses by the method of Bagley (*40*), but the Rabinowitsch correction for non-Newtonian effects was not used. The median fiber lengths were approximately 0.005 inch in each case, and the suspensions were so well mixed that no further fiber degradation occurred during testing.

Apanel (*37*) and Shelton (*38*) studied the effect of repeated capillary extrusion on the viscosity and fiber length distributions of 15 v/o ⅛-inch glass fibers in silicone oils. Photomicrographs of a typical suspension taken after various runs (Figure 4) clearly show that breakage occurs.

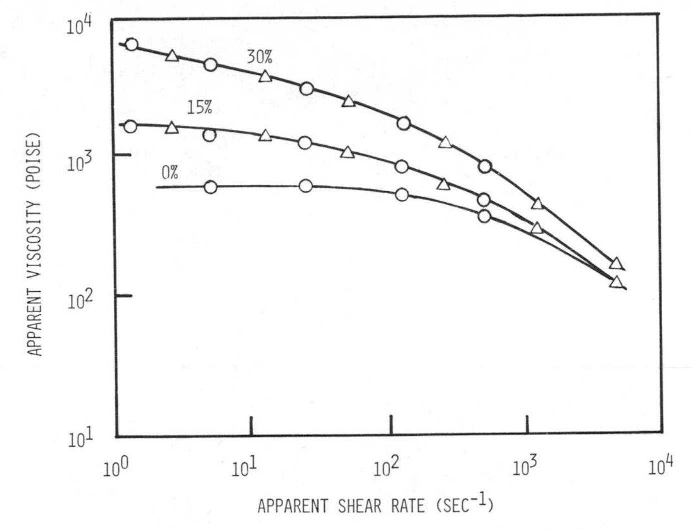

Figure 3. Apparent non-Newtonian viscosity of suspensions of glass fibers in 600-poise oil (○, ¼-inch capillary; △, ⅛-inch capillary)

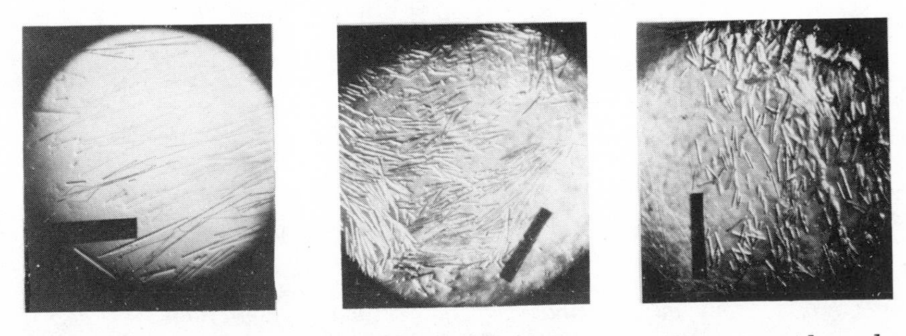

Figure 4. Photomicrographs of glass fibers after repeated extrusion through the capillary viscometer. Left to right, after 0, 10, and 20 runs. Dark bar is a bundle of fibers ⅛-inch long (37).

Figure 5 shows typical extrusion data. The curves for the 1- and 6-inch capillaries, when normalized, superimpose to yield one curve, indicating that fiber damage occurs in the capillary entry region and not during passage through the capillary. The extent of damage is higher at higher extrusion rates, but a suspension which showed no measurable degradation after multiple passes at low shear rates showed a considerable drop in viscosity when tested at high rates (Figure 5). Therefore, a drop in extrusion force may not be a sufficient test of fiber damage.

Roberts (*34, 35*) has completed study of the viscosity of mixed suspensions of various ratios of ⅛-inch glass fibers to 30-μ glass spheres, at a constant total solids volume fraction of 15% in the 600-poise silicone

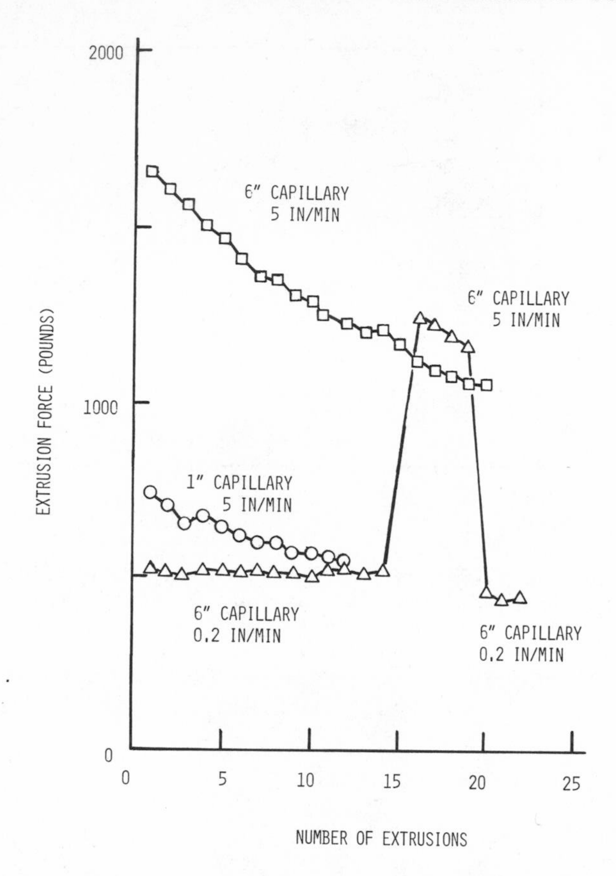

Figure 5. Effect of repeated extrusion on extrusion force for suspensions of 15 v/o glass fibers in silicone oil. Legends indicate capillary length and Instron crosshead speed (37).

oil. His data are shown in Figure 6, and plots of the Bagley end correction, *e* (*40*), *vs.* shear stress are shown in Figure 7. Even though these end corrections are large, the suspensions exhibited no die swell upon exiting from the capillary. They do exhibit a very strong Weissenberg rod climbing effect, as shown in Figure 8. The end corrections result from the large forces required to reorient the fiber mass as it enters the capillary and not from the usual effect of storage of energy in an elastic polymer network.

Research Needs

Despite the rapid growth in the use of short fiber reinforced composites, relatively few works on the rheology of concentrated fiber sus-

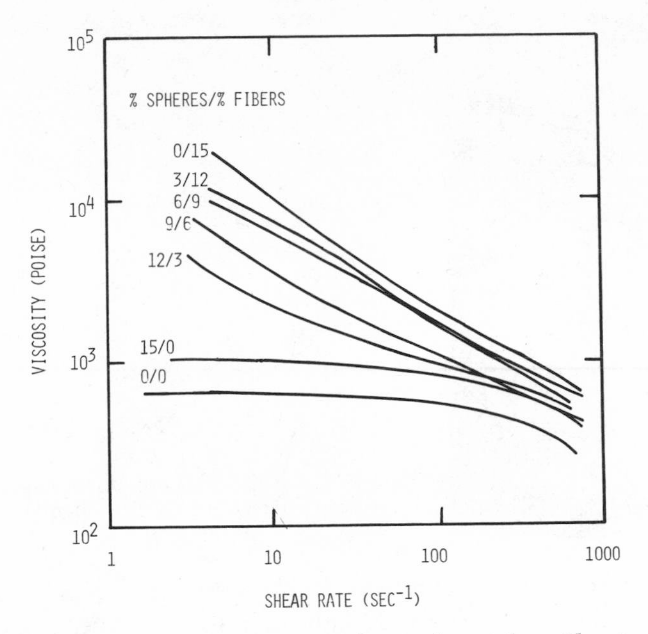

Figure 6. Viscosities of glass sphere/glass fiber/silicone oil suspensions (34).

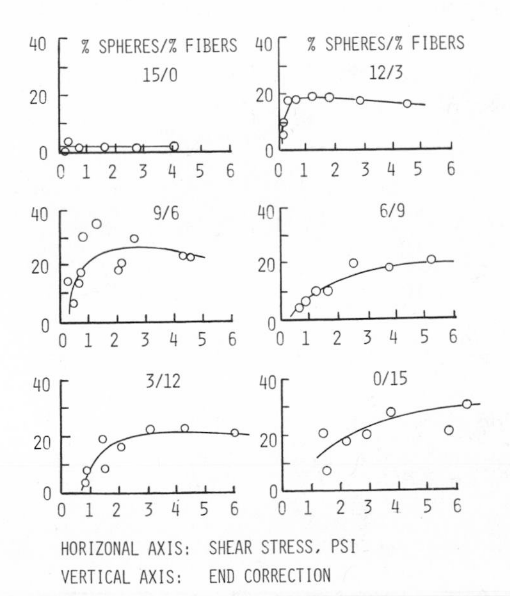

Figure 7. End corrections for glass sphere/glass fiber/silicone oil suspensions (34)

pensions have appeared. More experimental work is needed to elucidate the dependence of suspension rheology upon the properties of the components, the nature of their interface, the conditions under which they

Figure 8. Rod climbing during mixing of 3 v/o glass fibers, 12 v/o glass beads in silicone oil at 2 rpm (34)

are mixed, and the flow geometry. While volume fraction of the fiber phase is important, there is a clear indication that the fiber length distribution may play a key role in determining the magnitude of suspension viscosity. There is also a need for theoretical analysis or modeling of the flow properties of concentrated suspensions of fibers in which the mechanical interactions of the fibers are considered to be large. A complementary effort to predict the dependence of the fiber length distribution on mixing conditions and time is also needed.

Literature Cited

1. Maude, A., Whitmore, R., *Brit. J. Appl. Phys.* (1956) **7**, 98.
2. Maude, A., *Brit. J. Appl. Phys.* (1959) **10**, 371.
3. Vand, V., *J. Phys. Chem.* (1948) **52**, 287.
4. Goldman, A., Cox, R., Brenner, H., *Chem. Eng. Sci.* (1967) **22**, 637.
5. Segre, G., Silberberg, A., *J. Fluid Mech.* (1962) **14**, 136.
6. Karnis, A., Goldsmith, H., Mason, S. G., *Can. J. Chem. Eng.* (1966) **44**, 181.
7. Seshadri, V., Sutera, S., *J. Colloid Interface Sci.* (1968) **27**, 101.
8. Seshadri, V., Sutera, S., *Trans. Soc. Rheol.* (1970) **14**, 351.
9. Cox, R., Brenner, H., *Chem. Eng. Sci.* (1968) **23**, 147.
10. Attansio, A., Bernini, U., Galloppo, P., Segre, G., *Trans. Soc. Rheol.* (1972) **6**, 147.
11. Rutgers, I. R., *Rheol. Acta* (1962) **2**, 202, 305; (1963) **3**, 118.
12. Frankel, N. A., Acrivos, A., *Chem. Eng. Sci.* (1967) **22**, 847.
13. Brenner, H., *Ann. Rev. Fluid Mech.* (1970) **2.**
14. Eagland, D., Kay, M., *J. Colloid Interface Sci.* (1970) **34**, 249.
15. Woods, M. E., Krieger, I. M., *J. Colloid Interface Sci.* (1970) **34**, 91.
16. Papir, Y. S., Krieger, I. M., *J. Colloid Interface Sci.* (1970) **34**, 126.
17. Kao, S. V., D.Sc. Thesis, Washington University, St. Louis (1973).
18. Highgate, D. J., Whorlow, R. W., *Rheol. Acta* (1970) **9**, 569.

19. Agarwal, P. K., M.S. Thesis, Washington University, St. Louis (1971).
20. Nazem, F., D.Sc. Thesis, Washington University, St. Louis (1973).
21. Schmidt, L. R., M.S. Thesis, Washington University, St. Louis (1967).
22. Nazem, F., Hill, C. T., *Trans. Soc. Rheol.* (1974) **18,** 87.
23. Onogi, S., Matsumoto, T., Warashina, Y., *Trans. Soc. Rheol.* (1973) **17,** 175.
24. Stankoi, G. G., Trostyanskaya, E. B., Kazanski, Tu. N., Okorokov, V. V., Mikhasenok, Ya., *Soviet Plastics* (Sept. 1968), 47.
25. Thomas, D. P., Hagan, R. S., *Ann. Meetg. Reinforced Plastics Div., Soc. Plastics Ind., 1966.*
26. Newman, S., Trementozzi, Q. A., *J. Appl. Polymer Sci.* (1965) **9,** 3071.
27. Carter, L., Goddard, J., *NASA Rept.* **N67-30073** (1967).
28. Mills, N., *J. Appl. Polymer Sci.* (1971) **15,** 2791.
29. Bell, J., *J. Composite Material* (1969) **3,** 244.
30. Ziegel, K. D., *J. Colloid Interface Sci.* (1970) **34,** 185.
31. Takano, M., "Flow Orientation of Short Fibers in Rectangular Channels," Report #HPC-70-116. Monsanto/Washington University Association, February 1974.
32. Takano, M., unpublished data.
33. Karnis, A., Goldsmith, H. L., Mason, S. G., *J. Colloid Interface Sci.* (1966) **22,** 531.
34. Roberts, K. D., M.S. Thesis, Washington University, St. Louis (1973).
35. Roberts, K. D., Hill, C. T., *Ann. Tech. Conf., Soc. Plastics Engrs., Montreal, May 1973.*
36. Brodnyan, J. G., *Trans. Soc. Rheol.* (1959) **3,** 61.
37. Apanel, G., Undergraduate Research Report, Washington University, St. Louis (1971).
38. Shelton, R. D., Undergraduate Research Report, Washington University, St. Louis (1971).
39. Maschmeyer, R. O., D.Sc. Thesis, Washington University, St. Louis (1974).
40. Bagley, E. B., Schreiber, H. P., "Rheology," Vol. V, F. R. Eirich, Ed., Academic, New York, 1969.

RECEIVED October 11, 1973. Part of this work conducted under the Monsanto/Washington University Association sponsored by the Advanced Research Projects Agency, Department of Defense, Office of Naval Research contract N00014-67-C-0218 (formerly N00014-66-C-0045). Other portions supported by National Science Foundation grant No. GH34594.

11

Compounding of Fillers in Motionless Mixers

NICK R. SCHOTT, STEPHEN A. ORROTH, and ARUNKUMAR PATEL

Department of Plastics Technology, Lowell Technological Institute, Lowell, Mass. 01854.

A powdered polypropylene resin and a talc filler were dry blended in a two-port motionless mixer. Two types of blends were made using 22 and 40% talc. Blending is characterized as simple mixing in which the particles of the two components are spatially rearranged into a more random distribution without reducing the ultimate particle size. Visual examination showed that the degree of uniformity increased in the first four passes. Tensile and impact properties were measured for molded samples made from the various blends. Results indicate that the values become random within the first few passes. The values of the dry blended material fell below those of a milled sample which was used as an ideal.

The compounding of fillers, reinforcements, colorants, and other additives is a major processing operation in fabricating plastic resins and products. Powdered additives are particularly difficult to compound because they require special handling (*1*). As discussed by Carr (*2*), the behavior of solids in bulk handling depends on many factors which may be classified into two groups. The first includes properties which are determined by individual particles. Some of these are particle size and shape, surface area, density, hardness, and surface activity. A second major group of properties comprises those of particles en masse. Some of these are bulk density, porosity, angle of repose, compressibility, cohesion, dispersibility, and particle size distribution. To assume a uniform distribution of components the materials must be melt compounded or dry blended. Operations are usually done in either batch or continuous form on intensive mixing machines where a powder is milled into a resin melt or in a dry blend form as in a ribbon blender. Selection of mixer type is governed by the condition of the starting material (*3*). Plastics are highly viscous materials (10^5 to 10^6 cp) which are commonly worked on in screw extruders, kneaders, internal mixers (Banbury), and roll and pug mills. Most of the mixers above are the batch type. Since batch processes are more costly, continuous processes are more desirable for cost and labor savings.

This research was done to dry blend talc into a polypropylene resin by passing both components through a motionless mixer. It is essentially a simple mixing with no expected reduction in ultimate particle size. The degree of blending was evaluated by tensile, impact, and hardness tests on parts compression molded from the blends. The molded parts were also examined visually.

Experimental

Mixer. A two-port motionless mixer similar to that reported by Chen (*4*) was used. These mixing devices are right and left hand helices which are inserted in a tubular housing. The leading and trailing edges of two adjoining helices are welded together so that the edges are perpendicular to each other. A total of six right- and left-hand helices were welded into a 2.0-inch diameter stainless steel pipe. Each helix had a 1.5:1 l/d ratio, and the total length of the mixer was 20.5 inches.

A stream of solid particles flowing through the tube is split in two and twisted clockwise by the first helix. At the entrance to the second helix each split stream is split again and mixed with the split from the other stream. Simultaneously the mixed streams are twisted counterclockwise in this helix. The behavior of the stream particles repeats itself for each two helices in the mixer housing.

The experimental setup is shown in Figure 1. Various quantities of talc and polypropylene powder were packed into a 7-inch feeder with a maximum diameter of 3.75 inches. A sliding paper held the components in the feeder. When the paper was removed, material flowed into the mixer by gravity feed to a collector located at the outlet of the mixer. Particles fell into the collector which had the same dimensions as the feeder. Since only one mixer of six elements was available, the influence of additional elements was determined by passing the particles from the collector through the mixer several times. This action was assumed to be equivalent to a mixer with a multiple of six elements in which the material goes through in one pass. The assumption seems reasonable since the mixing action of each element is equivalent.

Resin and Filler. A general purpose polypropylene resin (Profax 6523, Hercules Inc.) was used. This resin is a homopolymer with a nominal melt flow of 4.0 and a general purpose stabilizer system. A Tyler standard screen analysis showed 77% −20 +40 mesh; the remaining 23% was −40 +100 mesh. The resin had a bulk density of 0.545 gram/cm^3. A powdered talc (Eastern Magnesia Talc Co.) was used as a filler. Screen analysis showed 92% was −40 +100 mesh; of the remainder 7% was +140 mesh while 1% was −140. The powdered talc had a bulk density of 0.58 gram/cm^3.

Mixing and Sample Molding. Two compositions of resin and filler were formulated containing 22 and 40 wt % talc. Batches of *ca.* 250-grams total weight were used. Each component was weighed and placed into a compartment of the feeder. The paper was removed, and the components passed through the mixer. A total of 10 batches of each composition were made. The polypropylene flowed more freely than the talc. An external vibrator was used to prevent bridging during the initial pass. In subsequent passes the flow of the blend increased as the resin and talc intermixed. After the required number of passes the contents were poured into a mold platen and compression molded into a 1/8 × 6 × 8-inch sheet at 380°–400°F. The following molding cycle was used: 2 min preheat at low pressure, 2 min heating at 6000 psig, and 3 min cooling at 400 psig. Compression moldings were also made of a 100% polypropylene sample and two milled batches which had the same composition as the dry blended mixtures.

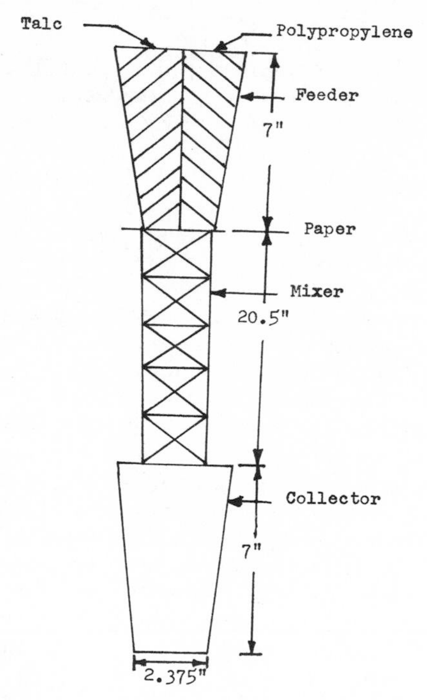

Figure 1. Schematic of mixing apparatus

Results and Discussion

Visual Examination. A visual examination of the molded sheets gives the first indication of the degree of mixing. The terminology of McKelvey (5) is used to describe each sample. Figures 2 and 3 show the molded sheets for the first one to four passes through the mixer. In the 22% samples of Figure 2 the talc concentrations are dark areas. The

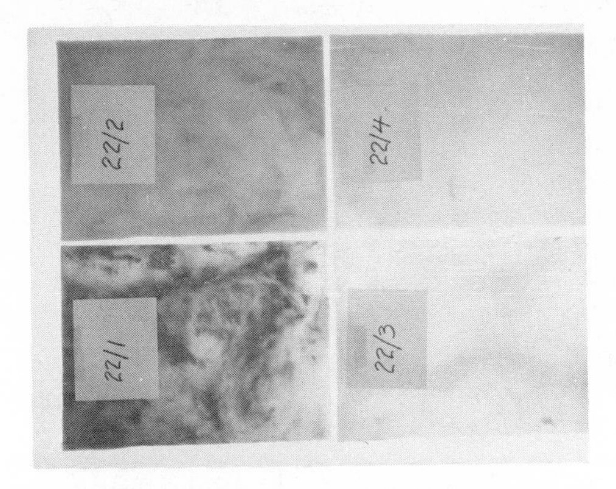

Figure 2. Compression-molded sheets (22% talc)

Figure 3. Compression-molded sheets (40% talc)

samples appear finely grained since the scale is large with respect to the size of the ultimate particles. The texture of each sample is indicated by the gross segregation of the components. In the first few passes this segregation is directly attributable to the flow properties of the components as discussed later. With one pass the distribution is very poor. Successive passes for each composition give added improvement. After five passes no additional improvement was seen visually, and the texture may be considered uniform. The 40% samples in Figure 3 show a similar trend. Again dark areas show the excess talc concentration. Sample 40/1 was poorly mixed because of the poor flow properties of the talc. The sample had very little structural integrity. In many places little or no resin was present which caused the sheet edges to crumble. The poor mixture was also observed in the surface finish. Samples became smoother

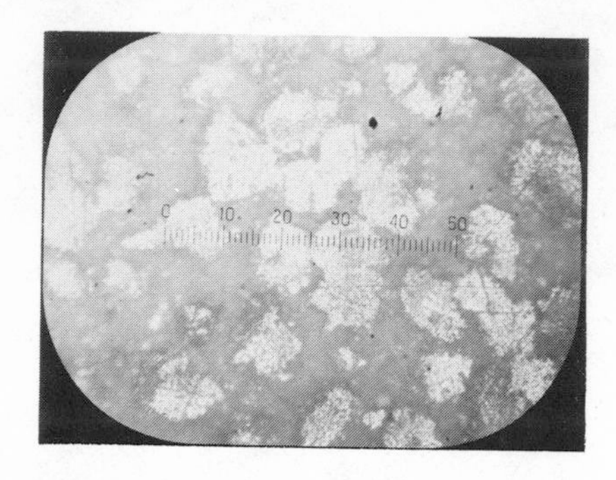

Figure 4. Surface of 40/2 sample at 50 × magnification

with each pass as more talc was wetted and coated by the resin. Figure 4 shows a 50 × magnification of the 40/2 mixture. The light areas are the resin particles surrounded by the much smaller unwetted talc particles.

Tensile Testing. Mechanical tests were done on test bars cut from the molded sheets. Table I shows the Young's modulus, ultimate strength, and percent elongation to break for the 22% composition as a function of the number of passes through the mixer. Property variation for the mixtures did not show significant trends as the number of passes was increased. Assuming that the milled material is ideal, Table I shows that the Young's modulus is not a significant criterion. The tensile strength and the elongation fell below the values for the milled sample. Table II shows the tensile properties of the 40% talc samples. Again these are compared with the milled sample. No values are reported for the first pass since the molded sheet was inferior (*see* Figure 3). The Young's modulus increases in the first three or four mixer passes whereafter variations become random. The tensile strength and elongation show similar behavior. The values for the dry mixed samples fall consistently below

Table I. Tensile Properties of 22% Talc-Filled Polypropylene

Mixer Passes	*Young's Modulus, psi × 10^5*	*Tensile Strength, psi*	*Elongation, %*
1	3.1	3860	4.0
2	3.05	3225	3.7
3	3.02	2450	2.6
4	3.62	3480	5.3
5	3.1	3170	4.6
6	4.1	3270	3.9
7	5.0	3290	5.1
8	3.7	3320	4.4
9	3.4	3320	5.2
10	3.6	3290	4.2
milled	3.4	4660	16.3

Table II. Tensile Properties of 40% Talc-Filled Polypropylene

Mixer Passes	*Young's Modulus, psi × 10^5*	*Tensile Strength, psi*	*Elongation, %*
2	2.84	1344	0.7
3	3.66	1523	0.6
4	4.03	2360	2.1
5	3.94	2400	1.5
6	3.58	2240	2.3
7	3.86	2420	2.0
8	4.42	2540	2.2
9	4.66	2400	2.2
10	4.47	2490	1.8
milled	6.11	4230	3.0

those for the milled sample. A comparison of Tables I and II indicates that filler loading at the 40% level affects the separate tensile test parameters and the number of mixer passes. Consistent with the visual observation that the degree of mixing for the 40% talc is poor at low passes, one observes that the samples become more uniform and the properties in Table II appear to stabilize after three or four passes. The observation was also observed in the flow of the material through the mixer. It took 3 or 4 passes before the blend flowed smoothly since large concentrations of talc disrupted flow.

Table III. Impact Testing for 22% Talc Formulation

	Impact Strength, ft-lb/inch	
Mixer Passes	*Unnotched*	*Notched*
1	4.89	0.84
2	2.43	0.88
3	1.32	0.88
4	2.60	0.86
5	2.60	0.87
6	2.62	0.90
7	2.80	0.94
8	2.32	0.83
9	2.23	0.85
10	3.44	0.89
milled	7.27	1.09

Impact Testing. Izod impact tests were done for notched and unnotched samples. Table III lists the values for the 22% composition. The unnotched impact values decrease in the first three passes where the mixing is poor. The high value in the first pass may be attributed to a resin-rich region since a 100% polypropylene sample gave 5.74 ft-lb/inch.

Such gross property variations are minimized as the degree of mixing improves. If enough tests were done, statistics would show that the standard deviation and the variance of a mixture property would decrease as the degree of mixing increases. The values vary randomly but are less than half the milled value (7.27 ft-lb/inch). Greater differences can be seen between the notched and unnotched samples. Table III shows consistent values between .83 and .90 ft-lb/inch for all passes. These indicate that the notch sensitivity is constant for the degree of mixing achieved in the first 10 passes. Similar behavior is obtained for the 40% talc mixture (Table IV). Here the notched samples also showed impact values which were constant with increased passes; also the additional talc loading gave no improvement. For the 40% talc unnotched samples the impact values varied randomly and fell below the milled value by a factor of 0.5.

Table IV. Impact Testing for 40% Talc Formulation

	Impact Strength, ft-lb/inch	
Mixer Passes	*Unnotched*	*Notched*
2	1.05	0.80
3	0.92	0.85
4	1.33	0.87
5	1.40	0.89
6	1.52	0.90
7	1.68	0.91
8	1.82	0.93
9	1.59	0.94
10	1.40	0.91
milled	2.96	0.99

Flow Behavior of Materials. Processing characteristics of a material depend on its bulk as well as its particulate properties. The flow rate of each component through the mixer was measured for gravity flow. The polypropylene resin flowed freely at 17.4 lb/min for a fully flooded (non-starved) condition. The talc, on the other hand, showed no tendency to flow by gravity. This behavior is caused by the material's particle size, size distribution, and shape. Lieberman (*6*) reports that the power required to maintain flow through ducts can be reduced by including a range of particle sizes rather than a single particle size. The screen analysis determination does not show that the resin had a significantly better size range than the talc. This suggests that the particle shape and material properties play a large influence. The tendency of the talc to pick up moisture and cake may explain its poor flow properties relative to polypropylene. Vigorous vibration of the mixer housing was necessary for the talc to flow. A maximum flow of 1.2 lb/min was observed. The degree of mixing is affected by these differences in flow rates for the two components. In the initial pass of a mixture the flow was erratic because of the hangup of the talc. Larger amounts of talc produced more erratic flow. Flow improves with each pass as the composition becomes more uniform. Thus the degree of mixing should show a nonlinear increase with each pass. Variations would arise from the flow properties of the components and their ratios in each mixture.

Conclusions and Recommendations

Motionless mixers can be used to prepare dry blends. Evaluation of molded sheets from these blends shows that dry blending, which represents simple mixing, is not as good as melt blending since it is done on a two-roll mill. The impact strength for notched samples is independent of the number of mixing passes and of the degree of talc content between 22 and 40%. The notched impact values came closest to the value for the milled samples. For Young's modulus and tensile strength, properties varied randomly as a function of mixer passes. However, visual examination indicated that the distribution of the components became more uniform in the first four passes.

Literature Cited

1. Schoengood, A. A., *SPE J.* (Feb. 1973) **29**, 21.
2. Carr Jr., R. L., *Chem. Eng.* (Oct. 13, 1969) **76**, 7.
3. Parker, N. H., *Chem. Eng.* (Aug. 30, 1965) **72**, 121.
4. Chen, S. J., Fan, L. T., *J. Food Sci.* (1971) **36**, 688.
5. McKelvey, J. M., "Polymer Processing," p. 301, Wiley, New York, 1962.
6. Lieberman, A., *Chem. Eng.* (March 27, 1967) **74**, 97.

RECEIVED October 11, 1973.

12

Compounding of Fillers

STAN JAKOPIN

Werner & Pfleiderer Corp., Waldwick, N. J. 07463

The several methods available to produce glass-reinforced thermoplastic materials vary greatly in their methods and results. Data obtained on pre-compounded glass-reinforced polypropylene show that physical properties vary significantly with various compounding techniques. Proper selection of compounding equipment and optimizing equipment parameters can substantially increase mechanical properties of the final product. Methods of selecting the most efficient equipment for a given compounding operation must take into consideration degree of shear required, temperature sensitivity, residence time distribution, and volume to be produced. When dealing with glass fibers, abrasion or corrosion of the compounding equipment plays a substantial role in economics. The simplicity and accuracy of the compounding process are also important.

The development of filled plastics has reached a point where properties of raw materials (filler and polymer) as well as final molding parameters have been thoroughly studied and substantial technical data have been published. On the other hand the technology involved in compounding fillers and polymers has not been well publicized, and this critical area directly influences both total economics and final end-product quality. The volume of precompounded plastic pellets used annually has been increasing at a remarkable rate. Selection of the proper compounding system, therefore, is a critical engineering function. Available processes must be carefully evaluated, taking into consideration end-product quality, degree of automation possible, and realistic production rates as well as versatility. A wrong decision here can cause an eventual commercial failure. A good evaluation can provide a competitive edge.

Today nearly all important polymers are available as filled precompounded pellets for injection molding and extrusion. Bearing in mind process economics, equipment availability, and preference, the decision as to what compounding technique and what compounding equipment to utilize often centers around the question of how much shear is required, how sensitive the compound and/or compounding ingredients are to temperature, and the volume to be compounded. Quality levels, space limitations, and throughput requirements must also be considered. The processor then can select either a separate or in-line system, whichever is most efficient for his needs.

This article reviews various methods of compounding fillers (reinforced and non-reinforced) into thermoplastic or thermoset materials. Before discussing the methods of compounding, we consider briefly the production requirements for compounding fillers.

The Compounding Production Processs

The equipment for compounding fillers and polymers must fulfill several requirements: (a) steady-state running conditions, (b) reproducibility of processing conditions, (c) ease of cleaning, and (d) versatility to adapt to new formulations. To achieve optimum material quality, the equipment should have:

(a) the ability to generate sufficiently high internal shear stresses to facilitate good dispersion of the additives

(b) the capability to expose each particle to short and equal stresses

(c) exact temperature control to regulate and minimize heat history

Compounding Methods. Basically, we can differentiate two types of compounding: (1) discontinuous system and (2) continuous system. The discontinuous system is fairly old and in most cases refers to Banbury (R) intensive mixers or roll mills. Throughputs for these systems range from 500 to 10,000 lbs/hr, and sizable investments are required. However, an efficient processing system will allow the compounder to operate economically at high volume. On the other hand, continuous compounding systems have capacities up to 7000 lbs/hr. Because of economics and the large volume requirements for filled plastics, continuous systems are usually preferred. For quality, continuous systems offer better uniformity of product with less batch-to-batch variation than a discontinuous system.

In most cases, the proper processing conditions for compounding can be filled adequately with single-screw compounding type extruders. The advances of screw design techniques and new devices that aid in localized and controlled introduction of shear are now available to a point where, in probably two out of three cases, single-screw extruders are adequate.

A relatively new approach in designing a single-screw compounding extruder is represented by a new medium shear type compounder called the Transfermix. This is a continuous, enforced order, stepless extensive, variable intensive mixer, a viscous heat exchanger, and vented extruder. The Transfermix consists of two opposite handed screws—a rotor turning inside the stator. In the narrow landed extensive mixing stages, the groove depth of the rotor decreases from maximum to minimum while, in the stator, it increases from minimum to maximum. Shear rate can be adjusted from 30 to 3,000 reciprocal seconds by speed variation and a running clearance adjustment.

For some compounding operations, twin-screw compounding systems are the most effective approach. If properly designed, twin-screw extruders provide maximum process controls, especially with respect to shear and stock temperature. Also with twin-screw compounders, large quantities of volatiles can be removed.

Types of Fillers. The many fillers used in the plastics industry today can be separated into two general categories: (a) reinforcing fillers, (b) non-reinforcing fillers. Both categories are applicable to either thermoplastic or thermoset resins. Chopped or roving glass is the most common type of reinforcing filler. Others commonly used are cotton, asbestos, Fybex, and so forth. The second category contains fillers such as clay, $CaCO_3$, talc, wood flour, and pigments. The type of filler (reinforcing or non-reinforcing) drastically affects the compounding process.

Compounding of non-reinforcing fillers usually requires the highest degree of dispersion—*e.g.*, carbon black in LDPE. Therefore, equipment should be able to generate high shear stresses to separate the agglomerates, particularly since these fillers usually have very small particle sizes. In compounding reinforcing fillers, the opposite approach is taken: low shear compounding must be used to prevent damage to the fillers. The main consideration is to wet the filler uniformly, devolatilize, and discharge.

Residence Time and Residence Time Distribution. The two essential elements in a successful continuous system are absolute control over residence time and residence time distribution. To minimize heat history, the residence time must be short and uniform during the entire process. Residence time for thermoset resins, for example, should not exceed 60 sec. This is also true for many heat- and shear-sensitive thermoplastic materials.

Residence time is primarily a function of machine design, screw RPM, and throughput. As in all continuous machines, extruders do not have an exactly defined residence time but rather a residence time spectrum. Uniformity of a continuous operation is illustrated by the type of spectrum. With screw machines in general we can differentiate four principal types of residence time spectra (Figure 1). The two identifying characteristics of a residence time distribution are the distance w between the points of inflection and the overall width b of the distribution curve.

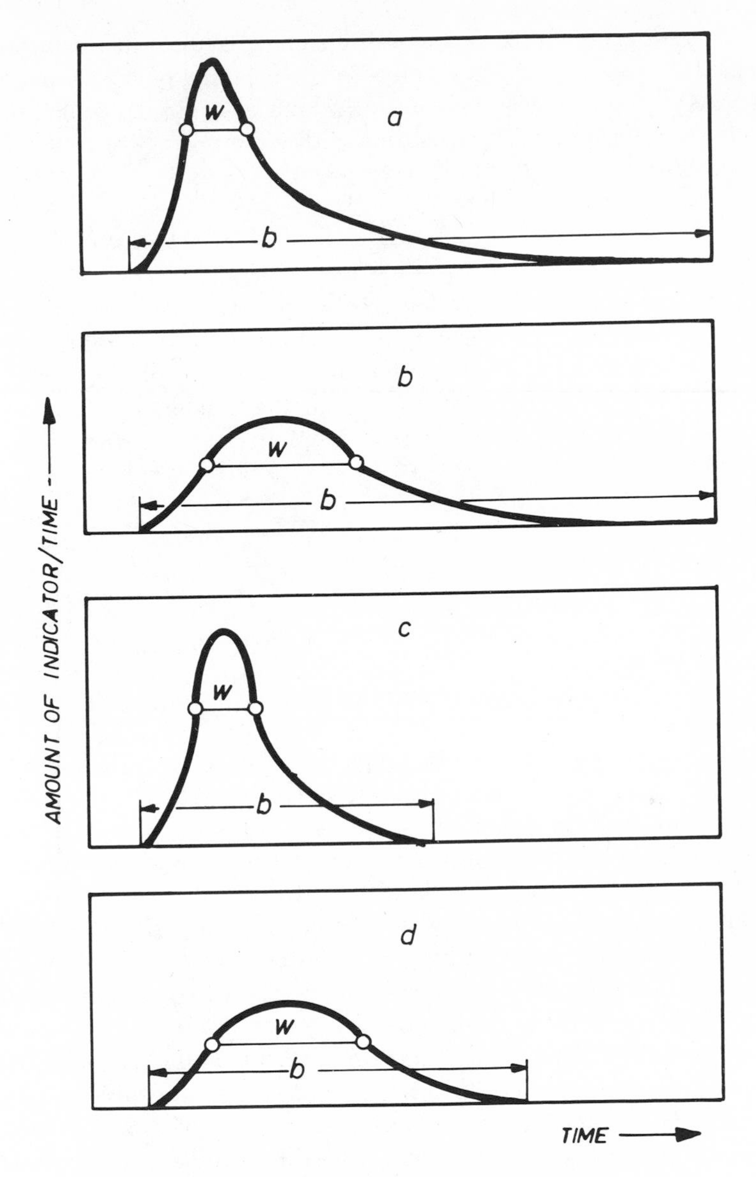

Figure 1. Four typical residence time curves for extruder processes (see text for discussion)

A small distance between the points of inflection indicates little back and forth mixing; a great distance indicates a greater longitudinal mixing. The overall width *b* of the distribution curve is influenced by so-called distribution tails which indicate the cleaning efficiency of the machine. Long distribution tails indicate poor self-cleaning efficiency.

The best characterization of self-cleaning is the self-cleaning time, s, defined as: $s = b - w$. As s increases, the cleaning efficiency decreases. Using average residence time, we can obtain a similar dimensionless value for self-cleaning characterization with various processes and average residence times:

$$\frac{s}{\bar{t}} = \frac{b - w}{\bar{t}}$$

Extruders will always have a value greater than 1. Like the term s, as the value increases, the self-cleaning characteristics deteriorate. Average residence time is defined as the time in which half of the particles in the residence time spectrum pass through the machine. When average residence time cannot be determined from a residence time spectrum, it can be calculated by:

$$\bar{t} = \frac{V \cdot \varepsilon}{\phi_v}$$

where V = free volume

ϵ = degree of fill

ϕ = volumetric flow per unit time

Curve a in Figure 1 is typical for a machine with little longitudinal mixing and poor self-cleaning characteristics. Curve b is typical for a machine with greater longitudinal mixing but still poor self-cleaning characteristics. These two curves are typical for single-screw extruders and for twin-screw extruders without a sealing profile.

Curves c and d show only short residence time tails which indicate good self-cleaning characteristics. The latter curves vary only in the amount of back and forth mixing and are typical of twin-screw extruders with a sealing profile. In machines with good self-cleaning, no particles remain excessively long in the unit where they might be subjected to severe heat. There are no dead corners where material could accumulate. These are critical factors, especially when processing thermoset materials. A good example of an extruder with a sealing profile is the twin-screw intermeshing and co-rotating compounder. The working principle of this unit is described below.

Twin-Screw Intermeshing and Co-Rotating Compounder. The processing section consists of two intermeshing screws, rotating in the same direction and at the same speed in the barrel. The screws are self-cleaning and wipe each other with a minimum clearance. Because of this sealing profile, dead spaces where material degradation could occur are minimized, and an even torque distribution is assured (Figure 2).

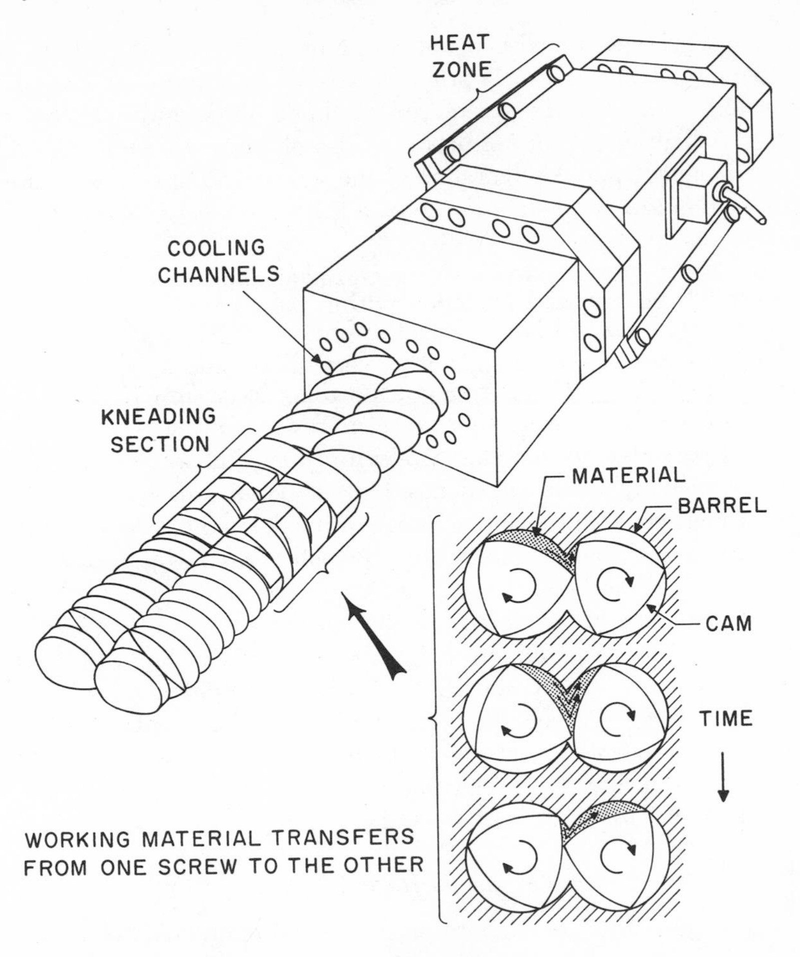

Figure 2. Twin-screw compounding extruder with co-rotating intermeshing screws. The screws are self-cleaning and wipe each other with a minimum clearance.

Screws and barrels are built up in a building block principle. The screw elements consist of different lengths and pitches and special kneading elements of various widths which are interchangeable. The screw elements and kneading blocks are secured on the shaft by a key. The screw elements are held on the shaft by the screw tip. By varying the screw elements and kneading blocks, the screw configuration can be tailored to the shear intensity required by the specific material. The screw barrel consists of individual barrel sections. When processing thermoplastic materials, the barrel before the discharge is usually the vented one where volatile constituents can be removed from the melt. Every barrel has cooling cores, so that close temperature control can be obtained with water or oil cooling. In many cases, heating is done electrically. This building block principle makes it possible to design the processing section exactly as required to obtain optimum processing conditions. A high degree of versatility is also obtained through the ability to vary the length and configuration of the screws and kneading elements. Thus, the required shear stresses can be adjusted to meet processing needs. The processing features are:

(a) The energy required to melt and homogenize the resin with the additives can be created by friction within the machine. This results in excellent dispersion and homogenization (ratio $w{:}b$ is very high).

(b) Energy can be created in a very short time and a very short machine length by high energy input kneading elements.

All this results in a very short average residence time—in most cases, below 30 seconds. Self-wiping and self-cleaning characteristics of the screw geometry prevent any deposition. Even in long-term operation, uniform melt conveyance and uniform product quality are maintained. These processing characteristics are especially critical for heat-sensitive materials where successful compounding depends on controlling the material temperature accurately while ensuring that all particles are exposed to a preset temperature for the same time. Since they fulfill these requirements, continuous high intensity compounders with small free volume and relatively short residence time have gained wide acceptance. The techniques used to compound fiber glass will illustrate the considerations involved in selecting compounding equipment and the influence of various equipment on final properties.

Glass Fiber Reinforced Thermoplastic Polymers

There are several methods to produce glass-reinforced thermoplastic materials, and they vary greatly in their techniques and results. Data obtained on precompounded glass-reinforced polypropylene show that physical properties vary significantly with various compounding techniques. Proper selection of compounding equipment and optimizing equipment parameters can substantially increase mechanical properties of the final product. The simplicity and accuracy of the compounding process are also important. When dealing with glass fibers, abrasion or corrosion of the compounding equipment also play a substantial role in total economics.

The properties of the composite are of course influenced by the glass fiber concentration, the strength of the fibers, and the effectiveness of the sizing agent. In the compounding operation, however, the quality of the reinforced polymer is directly affected by (a) the length of the glass fibers in the end product, (b) the uniformity of the glass distribution in the polymer, and (c) uniform wetting of the glass fibers by the plastic melt.

There are two basic ways of making a glass-reinforced thermoplastic part:

(1) material can be fed directly into an injection molding or extruding machine as a preblend made prior to molding, or

(2) Material can be purchased as pre-compounded pellets, which can be fed into an injection molding or extruding machine.

Single-Screw Extruders. The most commonly used system for compounding glass reinforced thermoplastics in single-screw extruders involves the use of chopped glass fibers which are preblended with polymer and fed into the extruder. The glass is conveyed together with the polymer through all three stages of single-screw extrusion: conveying, compression, and metering. During compression, when the polymer melts, the glass fibers are exposed to high shear stresses, and most of the reduction in glass fiber length takes place at this time. After melting, usually a venting section is used to remove the volatiles. The composite is then pushed through a die. Compounding in a single-screw extruder depends largely on head pressure which greatly influences the glass fiber length in the final product. Increasing head pressure damages the glass fibers which results in considerably decreased impact strength. Therefore, head pressure must be as low as possible.

Instead of using a premix, components can be fed separately into the extruder. Glass fiber can be in chopped or roving form; however, the screw should rotate fast enough to prevent build-up of material in the feed pocket—so-called starve feeding—otherwise the material might "arch" across the feed pocket and segregate because of vibrations. Glass damage during melting can be reduced by using a screw geometry which permits gradual melting by external heat and not strictly by mechanical energy input. Wear can be expected all along the screw section, particularly in the upstream section.

Compounding of glass fibers can also be done with the Transfermix system. The glass fiber length can be controlled to a certain extent by choosing the right shear rate for a particular polymer in order not to overwork the glass fibers. The machine is usually fed with a premix of glass fibers–polymer or by metering the components separately into the feed throat. Volatiles are removed through the vent section located downstream. Glass compounding with the Transfermix is similar to single-screw extruder compounding, but it offers improved versatility and control of fiber length.

Twin-Screw Extruders. A recent development in the production of glass fiber reinforced thermoplastic compounds involves the use of the intermeshing, co-rotating twin-screw compounder. For several good reasons, twin-screw compounding extruders have become more and more important for glass fiber composite production.

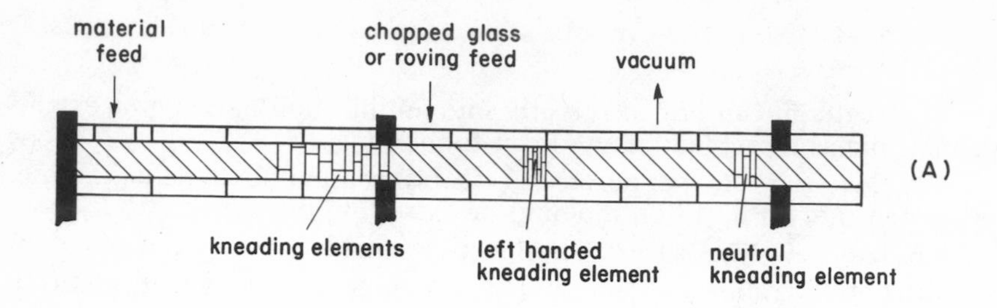

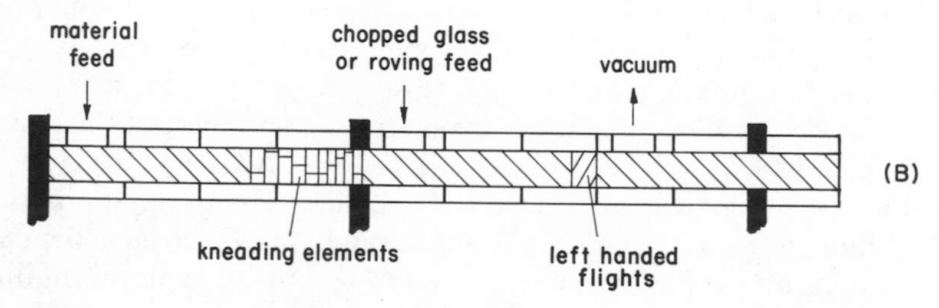

Figure 3. Typical screw configuration for compounding chopped or roving glass downstream into the melt

Variability of screw design is particularly important for producing glass fiber reinforced composites since it is possible to vary the required average glass fiber length, depending on polymer and the percentage of glass used. Polymers with high melt viscosities or high glass loadings (40% by weight or more) require milder screw configuration than polymers of low melt viscosity or low percent glass fiber (30% by weight or less). Figure 3 shows a typical screw configuration arrangement for two

Table I. Effect of Compounding on Mechanical

	Compounding Technique	
	Single-Screw Extruder	*Continuous Mixer*
Glass fiber, wt %	25	25
Type of fiber	1/8-inch chopped glass	1/4-inch chopped glass
Tensile strength, psi	6100	4700
Flex. modulus, m psi	580	460
Izod impact, ft lb/inch notched	1.45	0.7
Heat defl. temp., °F at 264 psi	264	153
% of Fibers smaller than 0.5 mm	—	—
Remarks		glass was fed into the feed section

different polymers: (a) low viscosity polymer (b) high viscosity polymer. The small kneading elements or screw elements with reversed flights can be used to determine the final fiber length distribution as well as the physical properties. The effect of a screw configuration on physical properties is shown in Table I. The properties are relative, and no specially treated polypropylene was used.

Similar to other processing methods, twin-screw intermeshing extruders can also be fed with a premixed polymer/glass blend, or components can be metered separately into the feed throat. Again, the equipment is usually starve-fed to prevent segregation. Since these compounders usually have high screw speed (200–300 rpm), there is no danger of being limited by the conveying capacity of the screws.

To prevent excess wear, twin-screw compounders are usually not fed with glass fibers into the feed port because of the abrasiveness of the glass. The wear to which the screws and the barrels are subjected in the plasticizing zone may be very severe. To prolong the equipment lifetime and to minimize the glass fiber length reduction during plasticizing, glass can be fed downstream into the melt through the degassing port or *via* a side feeder flanged onto the side of the barrel. The glass can be in roving or chopped form. The polymer is metered into the feed port in the conventional way and plasticized in the first section of the machine by a suitable screw configuration. If necessary, other additives such as flame retardants, pigments, plasticizers, or stabilizers, can be compounded thoroughly before the glass is fed. Also, any volatiles can be removed in that section.

Roving Process. The roving process uses continuous roving strands introduced into the melt through an open degassing adaptor, without

Properties of Fiber Glass-Filled Polypropylene

Compounding Technique			
Twin-Screw Compounder 1	*Twin-Screw Compounder 2*	*Twin-Screw Compounder 3*	*Twin-Screw Compounder 4*
25 roving	23 roving	25 roving	25 1/8-inch chopped glass
4900	5800	8000	8000
600	550	550	550
0.9	1.1	1.2	1.3
203	184	268	266
—	—	29	—
screw with very strong sections after addition of glass	moderate screw after addition of glass	mild screw after addition of glass (No. 14)	

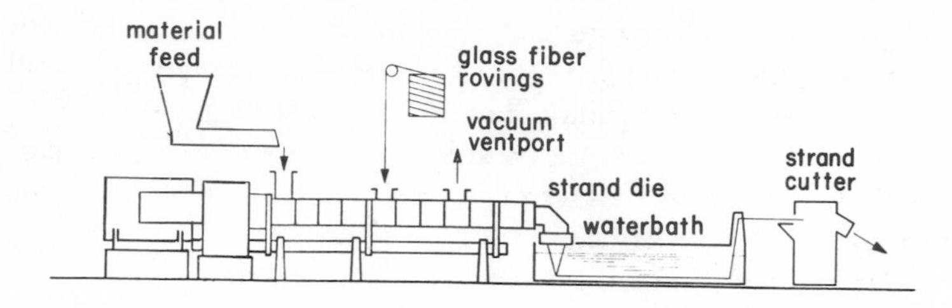

Figure 4. Typical arrangement for feeding continuous glass fiber rovings

Figure 5. Feeding glass fiber roving into a melt downstream through a vent opening of a twin-screw compounder

special metering devices. Strands are unwound from the cores continuously and pulled into the machine by the rotating screws. Since the intake of glass strands depends directly on screw rotation, the desired amount of glass can be controlled by the number of strands and the rpm of the screws. The amount of glass per hour of one strand can actually be plotted as a function of rpm.

If the feed rate of the polymer is known, the number of strands can be easily determined to match the right percentage of glass loading. Figures 4 and 5 show a typical arrangement for feeding continuous glass fiber rovings. The polymer feed rate can be readily varied because twin-screw extruders are generally starve-fed with no adjustment required in screw speed. In other words, screw speed can be varied independently of polymer feed rate, to adjust glass fiber feed. Roving spools can be set along the machine, or in larger production (20 or more strands introduced) the spools can be placed in the creels and pulled into the machine from one end.

The screw geometry in this part is designed so that the screw threads are only partially filled with polymer. This partial filling makes it possible for the screw to take up the glass fibers added at the feed point, while blockage of the glass fiber feed port by the plastic melt is avoided. The screw geometry downstream from the roving feed port is largely responsible for the fiber length and the homogeneity of the compound. Glass fiber is in endless form and must be chopped to a certain length in the machine and homogenized with polymer. To do this, proper screw geometry must be selected to meet these requirements. Rheological properties of polymer and loading of glass must be considered, and this can be done by installing kneading components or right- and left-hand screw elements in suitable combinations. Accuracy of the glass content is within $\pm 1.5\%$ or less as long as the polymer feed is constant and the unwinding rate of the roving remains constant.

Chopped Glass Process. Chopped rather than roving glass can also be fed downstream into the molten polymer, but metering equipment is required. As mentioned, the glass can be fed by gravity through a regular vent port or side fed with a delivery unit which is flanged to the barrel at a desired location.

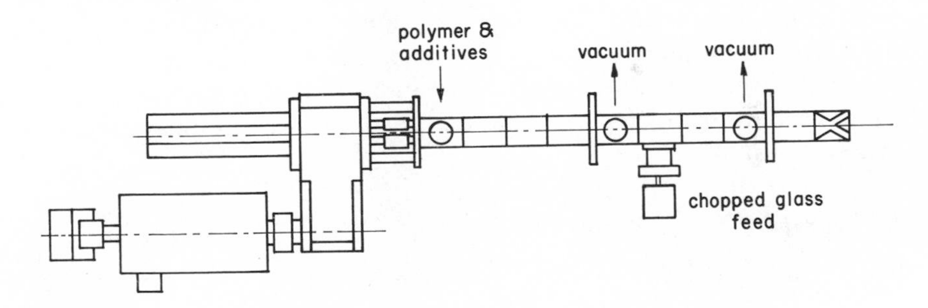

Figure 6. Arrangement for forced feeding of chopped glass downstream into the melt

Feeding by gravity has limitations as far as throughput is concerned. Using 2-inch extruders, feeding glass by gravity is satisfactory—70–100 lbs/hr can be fed easily except for polymers with sharp melting points (*e.g.*, polyamides or polyesters). When scaling up to a 3-inch extruder, the total throughput is limited by the glass feed. A maximum of 200–250 lbs/hr of glass can be fed by gravity. To overcome this limitation, a small side feeding unit can be attached to the side of the barrel. The glass is forced into the machine, as illustrated in Figure 6, and the throughput rate can be almost doubled from that of the gravity-fed rate.

With chopped glass, screw geometry can be quite different from that used with roving. Here, the glass fiber has to be only wetted by polymer, the volatiles removed and discharged through a die. Usually, straight conveying sections are used with various leads. For those polymers that are difficult to homogenize a small neutral kneading section is incorporated. Downstream feeding can also be used for other reinforcing or non-reinforcing fillers, particularly if they are abrasive.

Conclusion

To compound fillers properly, the equipment used must maintain steady-state running conditions and minimize batch to batch variations. It should be versatile and readily adapt to new and different formulations. The compounding process can adversely affect end-product properties. Two types of compounding methods are available: discontinuous and continuous. Continuous systems are preferred because they meet the throughput, economic, and rigid product quality requirements of most compounders. The essential elements in a continuous system are control over residence time and residence time distribution. Each process described here has its advantages and disadvantages. In many cases, compounding requirements can be met by a single-screw extruder. In other cases, where several processing functions must be performed in a single operation, twin-screw systems are essential. The intermeshing, twin-screw compounder with its interchangeable screw configurations offers the processor production versatility. It provides optimum shear and stock temperature control and devolatilizing capabilities needed for many of today's complex formulations.

Bibliography

1. Bernardo, A. C., "How to Get More from Glass-Fiber Reinforced HD,PE," *SPE J.* (Oct. 1970) **26**, 39-45.
2. Davis, J. H., "Fundamentals of Fiber-Filled Thermoplastics," *Plastics Polymers* (April 1971) 137-143.
3. Ross, Guenther, "Glasfaserverstaerktes Polypropylen," *Kunststoffe* (1970) **60**, 924-930.
4. Lees, J. K., "A Study of the Tensile Strength of Short Fiber Reinforced Plastics," *Polymer Eng. Sci.* (1968) **8** (3).
5. Schlich, W. R., Hagan, R. S., Thomas, J. R., Thomas, D. P., Musselman, K. A., "Critical Parameters for Direct Injection Molding of Glass-Fiber Thermoplastic Powder Blends," *SPE J.* (Feb. 1968) **24**, 43-53.
6. Cessna, L. C., Thomson, J. B., Hanna, R. D., "Chemically Coupled Glass-Reinforced Polypropylene," *SPE J.* (1969) **25**, 35-39.
7. Krautz, F. G., "Polypropylene Theory & Practice," Intensive Short Course, PIA, Inc., July 21-23, 1971, University of Massachusetts, Amherst.
8. Coneys, T. A., "Coupled Glass Reinforced Polypropylene—A New Structural Composite Material," Society of Automotive Engineers Meeting, Detroit, Jan. 13-17, 1969.

9. Ehrenstein, G. W., "Glasfaserverstaerkte Thermoplastiche Kunststoffe—Grenze und Anwendungsmoeglichkeiten," *Kunststoffe* (1970) **60,** 917-924.
10. Olmstead, B. S., "How Glass-Fiber Fillers Affect Injection Machines," *SPE J.* (Feb. 1970) **26,** 42-43.
11. Filbert, W. C., Jr., "Reinforced 66 Nylon—Molding Variables *vs.* Fiber Length *vs.* Physical Properties," *SPE J.* (Jan. 1969) **25,** 65-69.
12. Herrmann, H., "Schneiken Machinen," *in* "Der Verfahrenstechnik," Springer Verlag, Berlin, 1972.

RECEIVED October 11, 1973.

13

Filler Reinforcement of Plasticized Poly(vinyl chloride)

RUDOLPH D. DEANIN, RAYMOND O. NORMANDIN, and GHANSHYAM J. PATEL

Plastics Department, Lowell Technological Institute, Lowell, Mass. 01854

Plasticized poly(vinyl chloride) was compounded with 10–100 phr of 6 μ calcium carbonate, 4.5 μ calcined kaolin aluminum silicate clay, 0.18 μ thermal carbon black, 0.012 μ fumed silica, chrysotile asbestos of 10^{14} fibrils/gram and 200/1 aspect ratio, and ½-inch long chopped glass roving. All six increased indentation resistance and modulus and decreased low-temperature flexibility, extensibility, and abrasion resistance. Some fillers, at some concentrations, improved tensile strength, resilience, and hot strength. In terms of overall improvement of properties, silica was best, followed by glass, asbestos, carbon black, clay, and calcium carbonate in that order.

Incorporation of fibers into plastics generally produces tremendous improvements in rigidity, strength, toughness, dimensional stability, and often many other properties, forming the basis for the field of reinforced plastics (*1, 2*). Particulate fillers of fine particle size and high surface areas, especially carbon blacks, improve most properties of elastomers greatly and are therefore referred to as reinforcing fillers (*3, 4, 5*). In plastics, however, particulate fillers generally result in much less improvement in properties and are commonly considered low-cost extenders (*6, 7, 8, 9*). Actually their effects on processing and properties are much more important than their effects on volume cost, and thus they should be considered primarily from this point of view.

Plasticized poly(vinyl chloride) is intermediate between conventional plastics and elastomers and might be described as a thermoplastic elasto-plastic (*10, 11*). Fillers are frequently added to it, particularly in flooring applications (*8*). Consequently it provides a useful starting point for studying the effects of fillers on properties and for identifying those properties which are improved by fillers and might therefore be defined as reinforcements.

Experimental

Filled plasticized poly(vinyl chloride) samples of the following composition:

Stauffer SCC-686 poly(vinyl chloride)	100
dioctyl phthalate plasticizer	50
NL Tribase lead sulfate stabilizer	3
filler	X

were prepared by weighing the ingredients into a beaker, mixing with a spatula until homogeneous, and fusing and masticating 5 min at 160°C on a 6 inch × 12 inch differential-speed two-roll mill. Milled sheets were stacked and pressed 5 min at 166°C to form sheets ⅛ inch thick. These were cut into test samples and evaluated according to standard ASTM methods wherever possible. The fillers studied ranged from conventional non-reinforcing to reinforcing particulate fillers and on to short and long fiber reinforcements.

Calcium Carbonate. Calcium Carbonate Co. G-White, 6-μ median particle size, recommended for plasticized poly(vinyl chloride).

Clay. Harwick 5C calcined kaolin aluminum silicate, 4.5-μ average particle size, recommended for wire and cable insulation and other plastics compounding applications.

Carbon Black. Cabot Sterling FT thermal black, 0.18-μ diameter.

Silica. Cabot Cab-O-Sil M-5 fumed silica, 0.012-μ particle size. Loadings beyond 30 phr (parts per hundred of resin) were not possible with the present technique.

Asbestos. Union Carbide Calidria RG-144 chrysotile asbestos, 10^{14} fibrils per gram, 200/1 aspect ratio, recommended for reinforcing vinyls. Loadings beyond 60 phr were not possible with the present technique.

Glass. Pittsburgh Plate Glass ½-inch long chopped glass roving 3129. Loadings beyond 30 phr were not possible, and even these were not uniform, with the present technique.

The results are presented in Figures 1–8. In all figures, the following code is used:

Calcium carbonate - - - - - - - - - - CC
Clay — · — · — · — · — · — Cl
Carbon black — — — — — — CB
Silica ——— — ——— — ——— S
Asbestos ———— ——— ——— A
Glass ———————————— G

Results

Hardness. Plasticized vinyls are generally judged first by indentation hardness. This was measured as Shore Durometer hardness D-2 according to ASTM D-2240 (Figure 1). All six fillers increased hardness, ap-

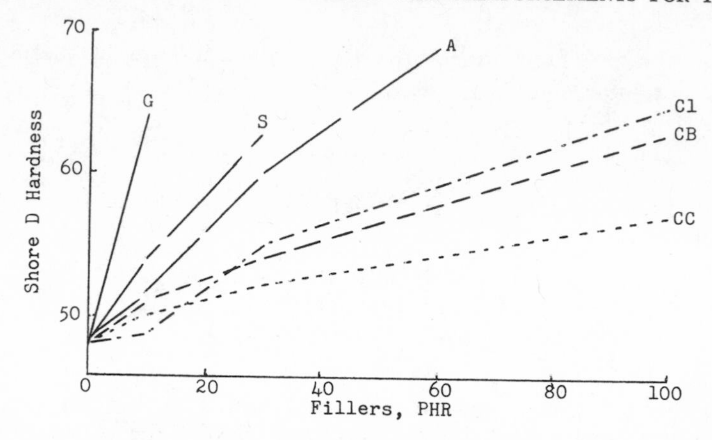

Figure 1. Effect of fillers on hardness

proximately in proportion to their concentration. Glass was the most effective, followed by silica, asbestos, clay, carbon black, and calcium carbonate. This property is important when furniture or pointed heels rest on vinyl flooring.

Modulus. Plasticized vinyls are generally formulated to a specified softness as measured by the tensile modulus at 100% elongation. This was measured according to ASTM D-638, using samples with flat section 2.32 inches long and a test speed of 2 inches per minute (Figure 2).

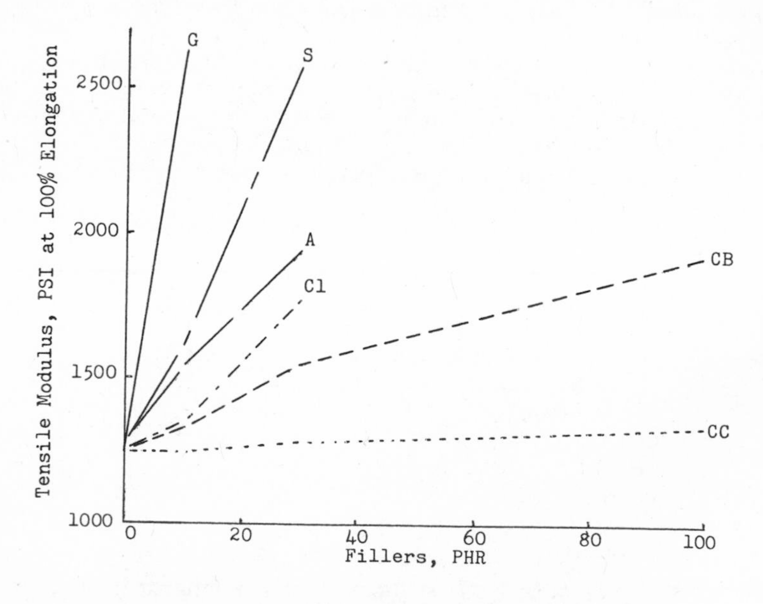

Figure 2. Effect of fillers on modulus

All six fillers increased modulus, approximately in proportion to their concentration. Glass was the most effective, followed by silica, asbestos, clay, carbon black, and calcium carbonate (this last had barely any effect). Unfortunately the stiffest compositions did not reach 100% elongation and thus could not be included in this plot. This stiffening effect may be explained by: (a) refusal of high modulus fibers to permit extension of the soft plastic matrix; (b) interference of high-modulus rigid filler surfaces with the normally free mobility of the soft plastic molecules; or (c) adsorption of plasticizer onto the high-surface filler particles, leaving much less plasticizer available to soften the plastic matrix. Practically, of course, filler loading offers an economical way to increase the stiffness of plasticized poly(vinyl chloride) to any desired level, for each individual application.

Low-Temperature Flexibility. The low-temperature flexibility of plasticized vinyls is generally judged by the Clash-Berg low-temperature torsional modulus test ASTM D-1043, and most commonly reported as T_F the temperature at which the material has a torsional modulus of 45,000 psi (Figure 3). All six fillers caused low-temperature stiffening of plasticized vinyl, approximately in proportion to their concentration. Glass was worst, followed by silica and asbestos, clay, carbon black, and calcium carbonate. The overall effect, 11° to 17°C at 100 phr of fillers, was serious enough to present a problem in low-temperature applications.

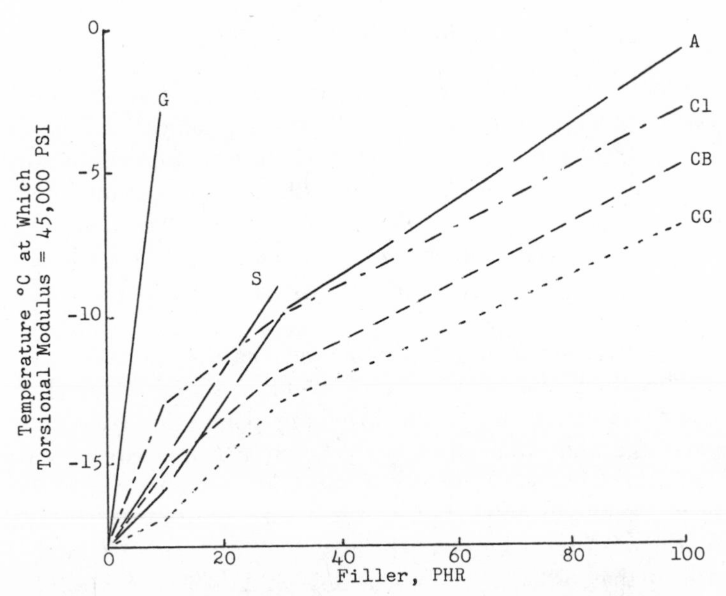

Figure 3. Effect of fillers on low-temperature flexibility

Tensile Strength. While most plasticized vinyl products are not generally used up to their ultimate tensile strength, this is the most common strength measurement made on such vinyl compositions. It was tested according to ASTM D-638 as described above (Figure 4). The improvement, caused by glass fibers was expected and should possibly have been higher with optimum processing. The improvement caused by finely powdered silica was unexpected and is the sort of beneficial

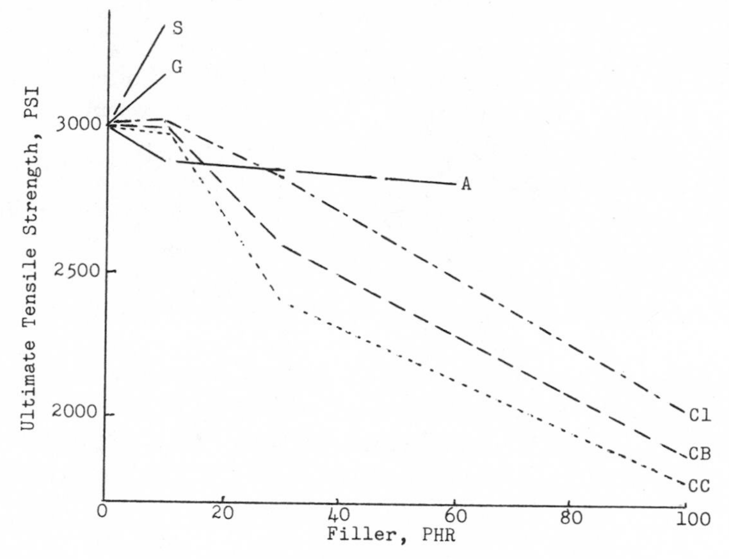

Figure 4. Effect of fillers on tensile strength

result normally labelled as reinforcement; with higher concentrations and suitable processing, it might be of major practical significance. All other fillers decreased tensile strength—a minor problem for asbestos, but increasingly serious for clay, carbon black, and calcium carbonate in that order. To prevent this sacrifice of strength, it would be of interest to study surface finishes which would increase interfacial bonding between the fillers and the plastic matrix, as has been done in many other reinforced plastics systems (*1, 2*).

Extensibility. Ultimate elongation was measured according to ASTM D-638 as described above (Figure 5). While plasticized vinyl products are not normally stretched to the breaking point, this property is generally measured and often erroneously taken as a measure of softness of the composition. In this study, all of the fillers lowered ultimate elongation, approximately in proportion to their concentration. Glass was most effective, followed by silica and asbestos, clay, carbon black, and calcium carbonate in that order. In applications where high extensibility is of practical importance, the use of fillers could thus be a problem.

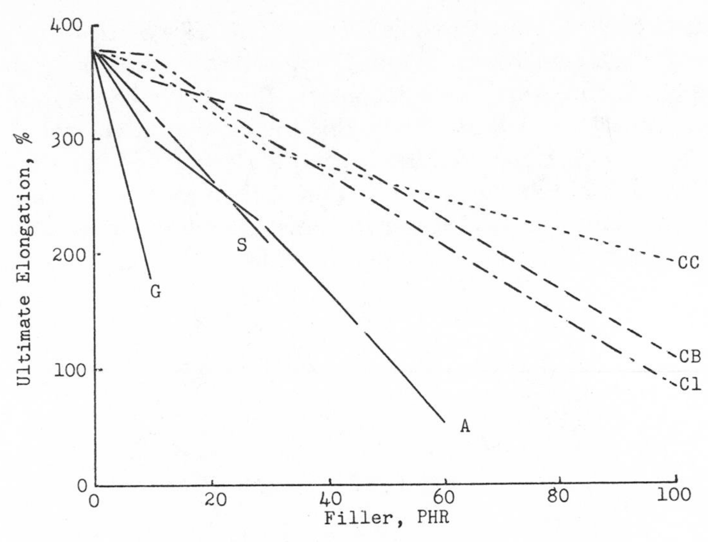

Figure 5. Effect of fillers on extensibility

Resilience. Rubbery resilience was measured by rebound according to ASTM D-2632, stacking ⅛-inch sheets one to five layers thick and recording the maximum rebound from each series (Figure 6). Low concentration of fillers produced some unexpected increase in resilience

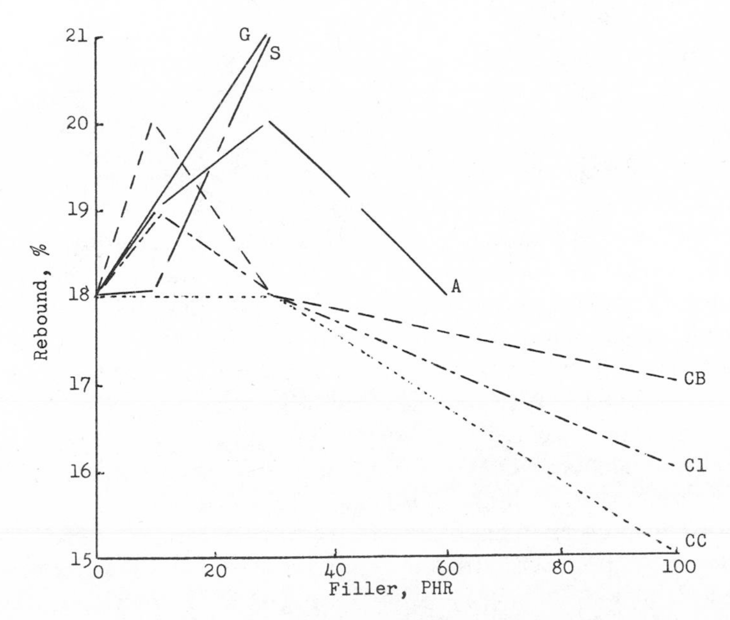

Figure 6. Effect of fillers on resilience

while high concentration produced a decrease. Theoretical explanation must await more intensive understanding of the mechanisms involved. Practically, all the vinyl formulations in this study had only 15–21% rebound, indicating high hysteresis; this means that at the speed of this test (about 6 mph) most of the mechanical energy of impact was absorbed by internal intermolecular friction and converted into heat. Thus such compositions are of interest for crash padding, shock absorption, packaging of delicate products, vibration damping, and noise damping (*12*).

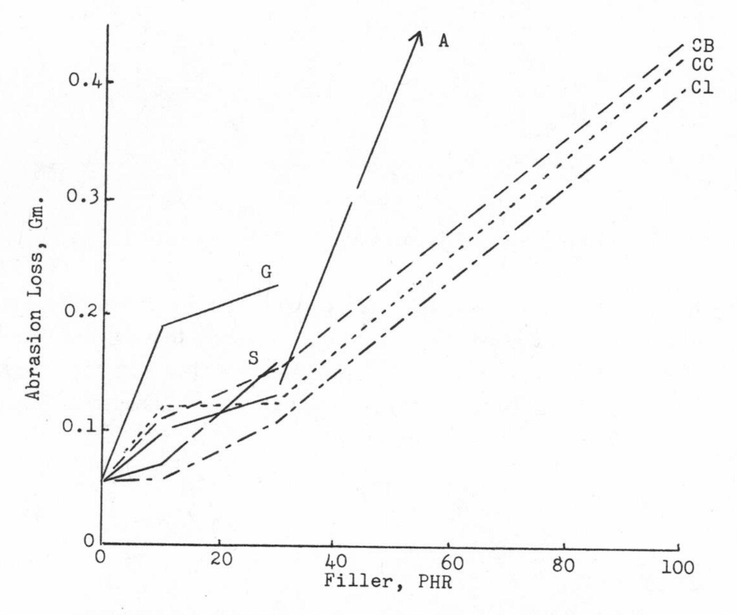

Figure 7. Effect of fillers on abrasion resistance

Abrasion Resistance. Abrasion was studied according to ASTM D-1044 using CS-17 wheels and a 1000-gram load; weight loss was measured after 5000 cycles (Figure 7). All six fillers increased abrasion loss, approximately in proportion to their concentration. Glass caused the greatest loss, followed by asbestos, with carbon black, calcium carbonate, and clay grouped closely together, and silica (hard to judge because of insufficient data). This loss of resistance to abrasion could be a problem in many applications. On the other hand, the exceptional durability of vinyl flooring, at still higher loadings, suggests that it may not be as serious as it appears. This discrepancy requires further study.

Hot Strength. Zero strength temperature (ZST) was measured by hanging 2 inch × 1 inch × ⅛ inch samples in a circulating air oven, increasing the temperature 1°C/minute, and observing the temperature at which each sample could no longer support its own weight and fell to the floor of the oven (Figure 8). Whereas unfilled plasticized vinyl failed at 200°C and low concentrations of fillers decreased this temperature somewhat, higher concentrations of four fillers produced marked improvement in hot strength, so that samples remained self-supporting up beyond the 285°C limit of the test. Silica and asbestos produced this effect even at 30 phr while carbon black and clay also produced it at 100 phr. The nature of the dip at low concentrations was mysterious and requires further study. The tremendous improvement at higher concentrations might arise from the effect of the fillers in increasing melt viscosity. It should be studied for its practical value in increasing the upper limit of the use temperatures for plasticized vinyl products.

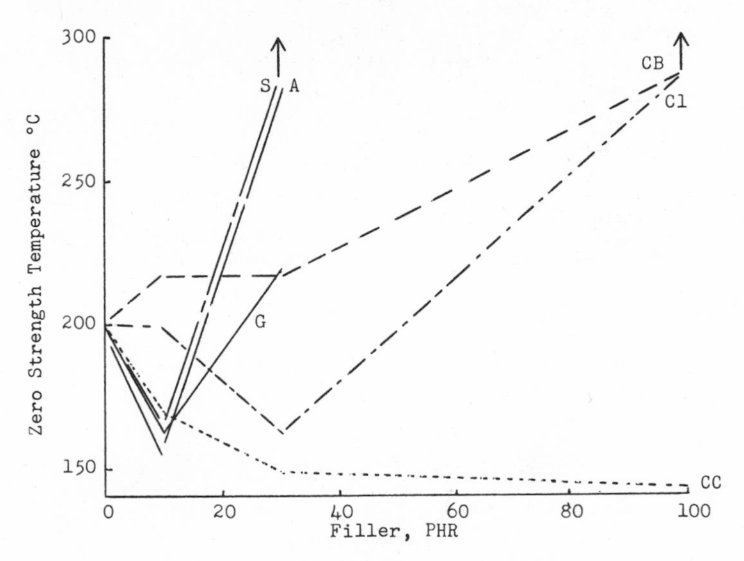

Figure 8. Effect of fillers on hot strength

Conclusions

All six fillers increased indentation resistance and modulus. Some fillers, at some concentrations, also improved tensile strength, resilience, and hot strength. All fillers decreased low-temperature flexibility, extensibility, and abrasion resistance. In terms of overall property improvement, silica was best, followed by glass, asbestos, carbon black, clay, and calcium carbonate. Optimum processing techniques to incorporate larger concentrations of glass fibers and silica should improve properties even further. Thus specific fillers can be chosen for specific improvements. Presumably, proper surface finishes on the fillers could increase their interfacial bonding to the plastic matrix and thus enhance strength and abrasion resistance.

Literature Cited

1. Lubin, G., "Handbook of Fiberglass and Advanced Plastics Composites," Van Nostrand Reinhold, New York, 1969.
2. *Mod. Plastics Encyc.* (1972) **49** (10A), 365-382.
3. Snyder, J. W., Leonard, M. H., in "Introduction to Rubber Technology," M. Morton, Ed., Chap. 8, Reinhold, New York, 1959.
4. Wolf, R. F., *Ibid.*, Chap. 9.
5. Kraus, G., "Reinforcement of Elastomers," Interscience, New York, 1965.
6. Boonstra, B. B., *Ibid.*, Chap. 16.
7. Brydson, J. A., "Plastics Materials," pp. 181-183, Van Nostrand, New York, 1966.
8. Sarvetnick, H. A., "Polyvinyl Chloride," pp. 107-113, 219-222, Van Nostrand Reinhold, New York, 1969.
9. *Mod. Plastics Encyc.* (1972) **49** (10A), 382-394.
10. Sarvetnick, H. A., "Polyvinyl Chloride," Chap. 5, Van Nostrand Reinhold, New York, 1969.
11. Brydson, J. A., *Ibid.*, Chap. 9.
12. Deanin, R. D., Shah, S. B., Kapasi, V. C., Pfister, D. H., *Amer. Chem. Soc., Div. Polym. Chem., Preprints* (1973) **14** (2), 861.

RECEIVED October 11, 1973. Based on a M.Sc. Thesis by G. J. Patel, Lowell Technological Institute.

14

The Effects of Moisture on the Properties of High Performance Structural Resins and Composites

C. E. BROWNING and J. M. WHITNEY

Air Force Materials Laboratory, Wright-Patterson Air Force Base, Ohio 45433

Graphite- and boron-fiber reinforced composites, as well as castings of current resin systems, were evaluated to determine the effects of moisture and/or high humidity on their physical properties and their room and elevated temperature mechanical properties. All of the neat resin castings absorbed moisture and swelled and showed a loss in elevated temperature tensile strength. All composite systems showed weight and thickness increases when subjected to high humidity. However, the effect of absorbed moisture on the elevated temperature mechanical properties is determined principally by the lay-up of the laminate and/or the test being applied. Thus, fiber controlled composite properties are relatively unaffected by absorbed moisture whereas matrix controlled properties are adversely affected. For both castings and composites the effects of moisture were reversible.

High performance structural composites have gone through several stages of development to where they are being readied for use in actual Air Force hardware. At this stage their performance during exposures to simulated aircraft enviroments must be evaluated. Important environmental factors include moisture (high humidity) and extremes in temperature. A major concern is retention of high temperature composite properties after exposure to high humidity.

This program was undertaken to determine the effects of high humidity on the mechanical and physical properties of high performance composites. Graphite- and boron-fiber reinforced composites as well as their more important associated cast resin systems were evaluated. Flexural and tensile properties were measured as a function of moisture, temperature, time of exposure to moisture, and number of exposure cycles. Quasi-isotropic laminates were studied because the critical design properties are the in-plane properties of multidirectional composites. Unidirectional properties were also obtained because certain ones (*e.g.*, flex) are very sensitive to matrix properties.

Table I. Materials Systems Evaluated

Cast Resins	*Composites*
	Boron 5505 (boron/ 2387)
Narmco 2387	HT-S/Erla-4617
Erl-2256	HT-S/Adx-516
Erla-4617	HT-S/P13N
Epon 828	A-S/X-2546
X-2546	HT-S/X-2546
	HM-S/X-2546
	HT-S/X-911
	HT-S/HT-epoxide
	A-S/Erla-4617

Table II. Fiber Properties

Fiber	*Modulus* (10^6 *psi*)	*Strength* (10^3 *psi*)	*Density* (*grams/cc*)
A-S	32.0	400	1.79
HT-S	38.0	380	1.75
HM-S	50.0	320	1.90
Boron	60.0	500	2.63

Experimental

The cast resins and reinforced composites evaluated are shown in Table I. Boron-reinforced composites were fabricated from Avco 5505 prepeg tape. Graphite fiber-reinforced composites utilized Hercules type A-S, HT-S, or HM-S graphite fibers. The properties of these reinforcements are given in Table II. Erla-4617, Erl-2256, and Epon 828 were evaluated because of their extensive use as laminating epoxy resin systems, particularly with graphite fibers. X-2546 is a newly developed, modified epoxy resin from Union Carbide having a high heat distortion temperature (485°F) and concurrent usefulness in composites at 350°F. Avco 2387 was evaluated because it is the resin system used in Avco 5505 boron/epoxy prepreg tape (the most widely used materials system in advanced composite developmental programs).

Cast resin and composite test specimens were subjected to the stepwise environmental exposure cycle shown in Table III. The cyclic exposure is representative of the severe environments an aircraft can see during service. Data were obtained at room temperature, 350°F after a 1-hr soak,

Table III. Stepwise Environmental Exposure Cycle

Step 1. Relative humidity = 95–100%. Temperature = 120°F. Time = 22½ hrs.

Step 2. Specimens are removed from chamber and placed under normal room conditions for 15 min.

Step 3. Temperature = −65°F. Time = 1 hr.

Step 4. Step 2 repeated.

Step 5. Temperature = 250°F. Time = ½ hr.

Step 6. Step 2 repeated.

and at 350°F (1-hr soak) after 2, 10, 15, and 30 exposure cycles. When only humidity aging was of concern, the humidity and temperature conditions shown in step 1 were used for 24 hrs. Aging times generally consisted of 15 or 30 days. 350°F is the maximum use temperature that most of these high performance composites will see in actual use and, therefore, is the test temperature of prime concern.

Boron and graphite composite test specimens were also subjected to water boil exposures of varying duration. Specimens were either boiled until their weight pick-ups were equivalent to their weight gains recorded during cycling—"equivalent water boil"—or until their weight pick-ups reached a constant value—"equilibrium water boil." A secondary factor of interest was whether or not water boil exposures could be used as quick and effective screening tests prior to cyclic exposures. Specimens were tested to determine if the same mechanical properties were found for two groups of specimens having the same weight gains but with one set having been cycled to the particular weight gain and the other set having been water boiled to the same weight gain. Table IV shows the short-beam shear strengths at room temperature and 350°F for both cycled and water boiled boron/epoxy specimens. At 350°F after 30 cycles (weight gain = 0.82%) and at 350°F after 11 hrs of water boil

Table IV. Comparison of the Shear Strengths of Water-Boiled and Cycled Specimens of Avco 5505 (Boron/Epoxy)

Test Conditions	*Shear Strength (10^3 psi)*	*Weight Gain (%)*	*% Retention of RT Strength*
Cyclic Exposures			
RT	13.3	—	—
350°F (dry)	6.0	—	45.1
350°F (after 30 cycles)	5.6	0.82	42.1
Water Boil Exposures			
RT	14.4	—	—
350°F (dry)	6.9	—	48.0
350°F (after 11 hrs water boil)	6.7	0.81	46.5

(weight gain = 0.81%) the percent retention of room temperature shear strength is approximately the same.

Flexural testing was done on 4.0 inch long × 0.5 inch wide specimens using the three-point loading method with a span-to-depth ratio of 32 to 1. Short-beam shear strengths were measured by the three-point loading method, using a span-to-depth ratio of 4 to 1. Test specimens were 1.00 inch long × 0.25 inch wide. Tabbed tensile test coupons were 6.0 inches long × 0.5 inch wide.

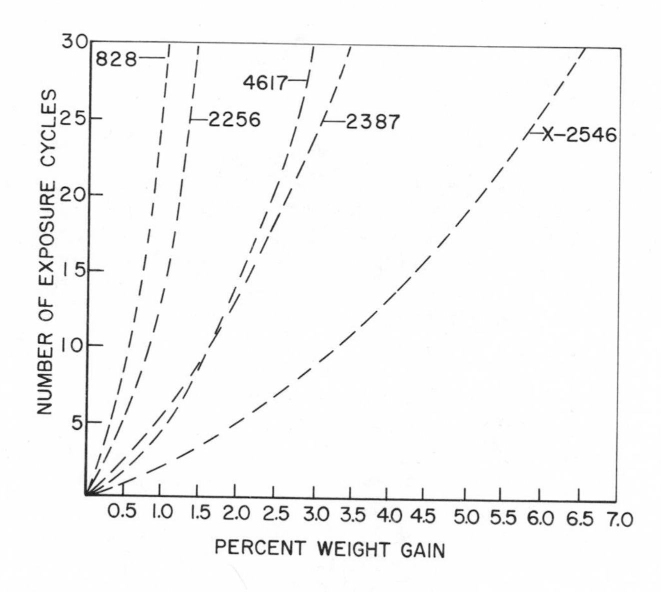

Figure 1. Effect of cyclic exposures on the weight gains of cast resin systems

Results and Discussion

Cast Resins. Weight gains from absorbed moisture for the cast resins are plotted as a function of the number of exposure cycles in Figure 1. The lowest weight gain after 30 cycles was about 1% by Epon 828 while the highest was about 6.5% by X-2546. The shapes of the curves indicate that none of the systems has reached equilibrium even after 30 cycles. Each system also showed concurrent thickness increases, varying from 1.3% for 828 to 2.8% for X-2546.

The effect of cyclic exposures on the tensile strengths of the cast epoxies is illustrated in Figure 2. The bar graph shows percent retention of the unexposed room temperature (RT) tensile strength as a function of the number of exposure cycles. Each system was tested near its proposed use temperature—*i.e.*, 2256 was tested at 250°F while the others were tested at 350°F. All systems showed large strength reductions caused by temperature (0 exposure cycles). After 30 exposure cycles none of the 350°F systems showed strength retentions greater than 15%. Erla-4617 fared so poorly that testing was dicontinued after 15 cycles. In a separate experiment, specimens of the 4617 system were ambient aged [RT and 50% relative humidity (RH)] for six months, after which their 350°F strength dropped from an unaged value of 5700 psi to an aged value of 1330 psi (77% reduction). These specimens were then dried at 250°F *in vacuo* to constant weight, resulting in the recovery of their original, unaged 350°F strength.

This reversibility aspect of the moisture degradation process is further illustrated in Figure 3 where the heat distortion temperatures (HDT) of 4617°C are plotted as a function of humidity aging. The HDT is the temperature at which a deflection of 0.02 inch occurs under a constant load. The solid line is the unaged control. The line furthest to the left

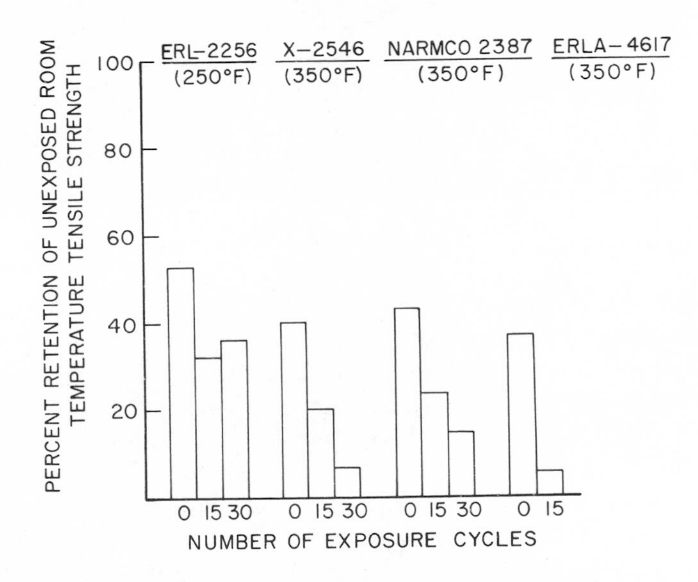

Figure 2. Effect of cyclic exposures on the tensile strengths of cast epoxy resins

was recorded after 20 days humid aging (step 1 conditions for 24 hrs), showing the HDT being lowered approximately 35°C; more importantly, the temperature at which the initial deflection occurs—an indication of the softening point of the resin and, in turn, the softening point of the composite—was lowered approximately 90°C. The corresponding moisture pick-up after 20 days aging was 3 1/2%. To illustrate the reversibility of the moisture absorption or plasticizing process, specimens which had been aged for 20 days were dried to constant weight, giving the dotted curve which is almost identical to the curve of the original, unexposed specimens.

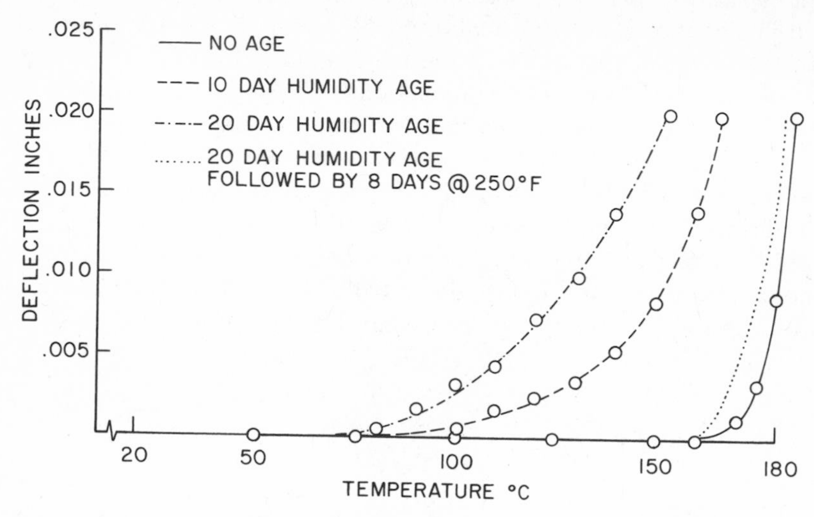

Figure 3. Effect of humid aging on the heat distortion temperature curves of Erla-4617 cast epoxy resin

Results and Discussion

Composites. The effect of cyclic exposures on the tensile strengths of quasi-isotropic HT-S graphite laminates is shown in Figure 4. Test temperatures and resin systems are indicated at the top of the figure. These are eight-ply laminates having ply orientations of 0°, +45°, −45°, 90°, 90°, −45°, +45°, 0°, with the test direction parallel to the outer 0° ply. 4617 and ADX-516 (an epoxy–polyarylsulfone) composites are so adversely affected by the temperature that plasticization by moisture could cause no further strength reductions. HT-S composites with P13N (a polyimide from Ciba-Geigy), X-2546, and X-911 (an epoxy–phenolic from Fiberite) show negligible 350°F strength losses as a function of cycling. P13N composites were also tested at 500°F with no strength reductions. Because of the ply orientations (fiber direction) and test direction, this particular property (quasi-isotropic tension) is fiber dominated (or controlled) to such an extent that substantial moisture absorption by the matrix does not cause significant strength reductions. Even the 4617 system which retained only 5% of its pure resin casting strength when tested at 350°F after 15 cycles, gave a quasi-isotropic graphite composite that retained 50% of its tensile strength at 350°F after 30 cycles.

The effects of moisture on the flexural or bending strengths of both unidirectional and quasi-isotropic graphite composites are illustrated in Figure 5. All tests were done at 350°F. Here, however, an equivalent water boil (*see* experimental) was used as an accelerated screening test. Flex or bending data show how the test method, fiber orientations, and resin HDT all influence the effects of absorbed moisture.

Several interesting comparisons can be made with the data in Figure 5. The quasi-isotropic tension in Figure 4 can be compared with the quasi-isotropic flex in Figure 5. The X-2546 and X-911 composites after being given equivalent water boils showed substantial reductions in their RT and 350°F bending strength because of absorbed moisture. The X-911 system which previously showed no tensile losses, lost over 40%

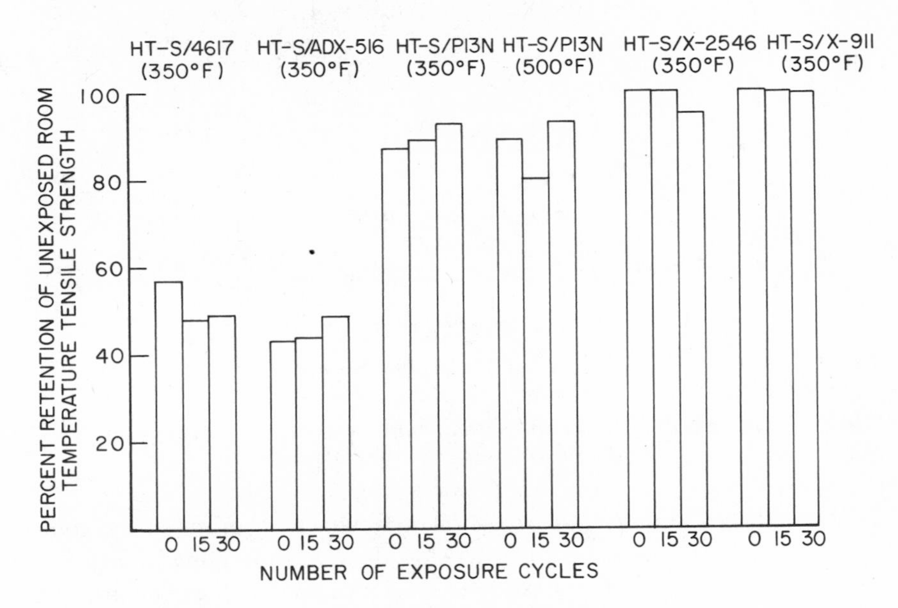

Figure 4. Effect of cyclic exposures on the tensile strengths of quasi-isotropic graphite composites

of its RT flex strength and 30% of its 350°F strength. Unidirectional X-2546 data show how moisture affects a matrix controlled property such as flex. There is a continual, stepwise decrease in the 350°F unidirectional flex strength with increasing water boil exposure time until, after 26 hrs, the strength retention was only about 50%. A comparison can also be made of fiber dominated *vs.* resin dominated lay-ups. Unidirectional 4617 composites with all of their fibers in the 0° direction are not nearly as adversely affected by the temperature as are the quasi-isotropic composites which have only two plies of eight oriented at 0°. Resin HDT can also be compared. The 4617 system which has a HDT of about 350°F gave a quasi-isotropic composite having a 350°F strength retention of less than 20% while X-2546 which has a +450°F HDT gave quasi-istotropic composites having 350°F strength retentions greater than 80%.

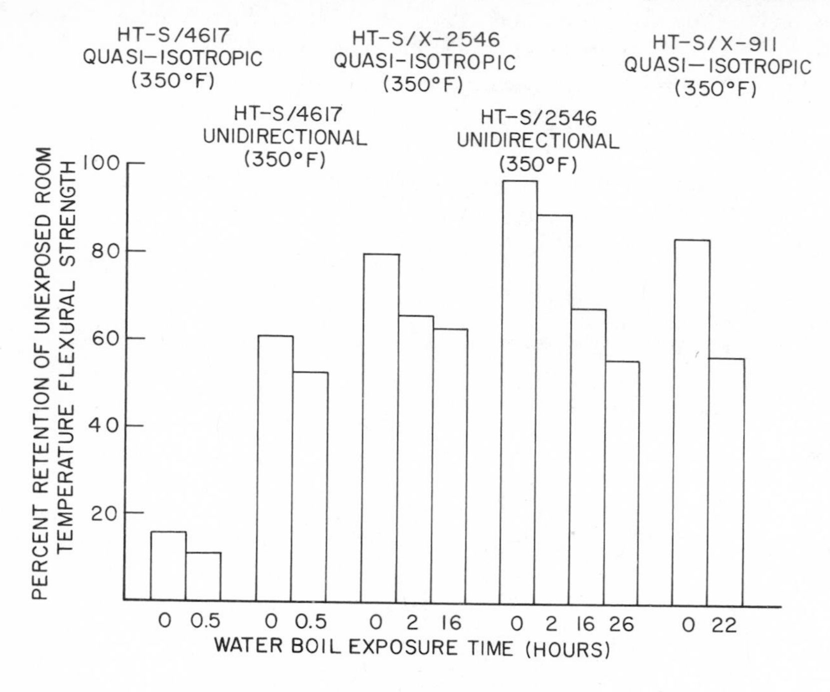

Figure 5. Effect of moisture (water-boil exposure) on the flexural strengths of graphite/epoxy composites

Strength retention in both tension and flex for boron/epoxy composites is shown in Figure 6. The tension data were obtained on quasi-isotropic coupons, and the flex data on unidirectional specimens. As shown, strength reductions are negligible at 350°F after 30 exposure cycles for the quasi-isotropic tension, even though the system had a weight pick-up of about 1%. In flexure, however, a drastic strength reduction of *ca.* 55% occurred because of absorbed moisture (equivalent water boil). Figure 7 shows the load–deflection curves from this same flex testing of the boron/epoxy specimens. If the 350°F "dry" curve is compared with the 350°F "wet" curve, there is a considerable loss in load-carrying ability with a simultaneous large increase in deflection caused by absorbed moisture. The test was actually stopped at the deflection shown without having broken any of the boron filaments. When tested at 350°F dry there is always filament breakage in the tension (bottom) face of the coupon. In other words, moisture absorption caused

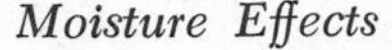

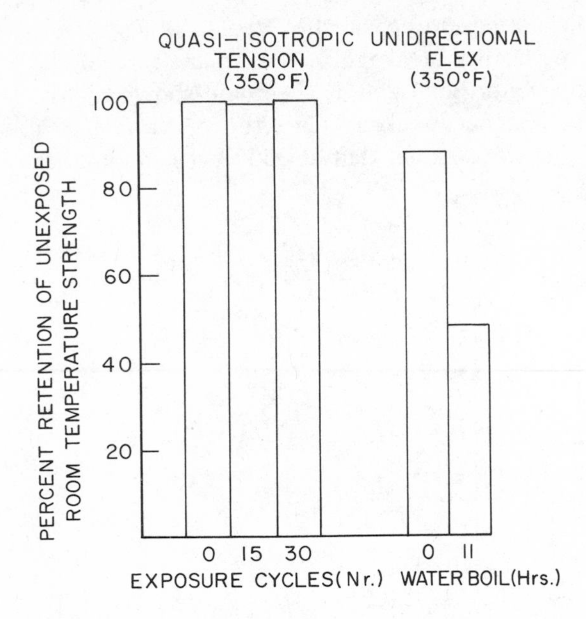

Figure 6. Effect of moisture on the tensile (quasi-isotropic) and flexural (unidirectional) strengths of boron/epoxy composites

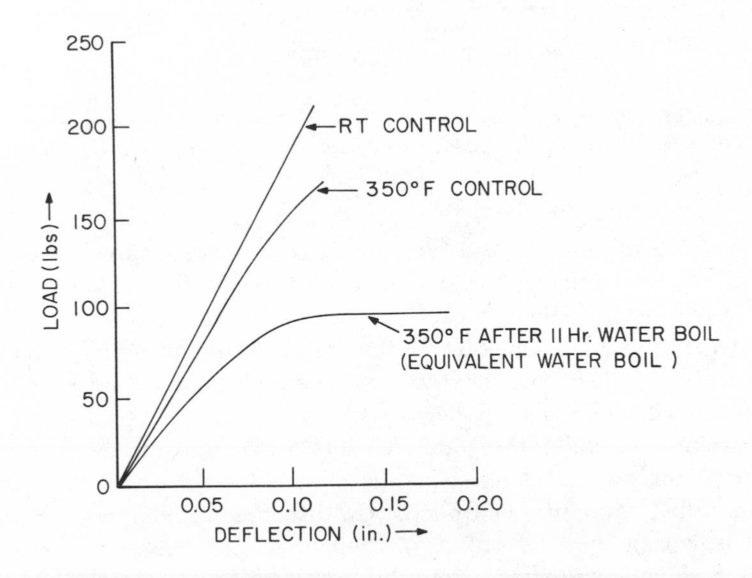

Figure 7. Load–deflection curves for unidirectional boron/epoxy composite test specimens

a different failure mode to occur; the "dry" 350°F control gave a sharp tensile failure with fiber breakage in the tensile face of the coupon; the wet (equivalent water boil) 350°F specimen showed only plastic deformation with no filament breakage (*i.e.*, the moisture has plasticized the epoxy matrix to such an extent that at 350°F it cannot efficiently transfer load from fiber to fiber).

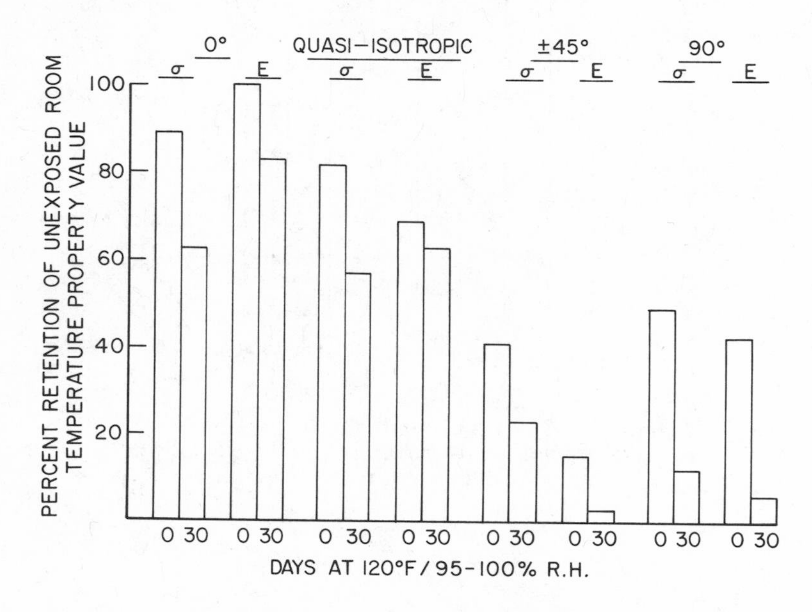

Figure 8. Effects of humid aging on the strengths (σ) and moduli (E) of A-S/Erla-4617 composites as a function of composite stacking sequence

The effects of moisture as a function of composite stacking sequence, lay-up or ply fiber orientations are illustrated in Figure 8. Shown is the percent retention of original, unexposed room temperature tensile strength (σ) and modulus (E) before and after 30 days humid aging (120°F/ 95–100% RH). The composite system is type A-S graphite and Erla-4617 epoxy resin with all testing done in tension at 300°F. As the graph is viewed from left-to-right, there are two sets of columns for the strengths and moduli for each stacking sequence shown—*i.e.*, the test goes from fiber controlled to matrix controlled. Unidirectional and quasi-isotropic specimens, with a large percentage of their fibers parallel to the direction of the test, show less ill effects from the humidity (even though they had substantial weight gains) than do the ±45° and 90° composites which have no 0° plies. The strength and modulus retention values for both the 0° and quasi-isotropic specimens were over 60% after 30 days humid aging while the $\pm 45°$ and 90° specimens showed less than 10% retention after equivalent aging. One disturbing result of this study was that quasi-isotropic specimens that had been humid aged for 30 days developed large cracks after setting in ambient conditions. Figure 9 is a photomicrograph (side view) of one of these cracked specimens, showing the crack running in the 90° plies.

Conclusions

The effect of absorbed moisture on the elevated temperature mechanical properties of composites is determined principally by the lay-up of the laminate and/or the test being applied—*i.e.*, the method by which load is introduced into the laminate. This means that a given type of laminate undergoing a specific method of elevated temperature mechanical testing may show no loss in the particular mechanical property being measured (at temperature) even though it has absorbed a significant amount of moisture. On the other hand, this same system, having a new lay-up, undergoing a different high temperature mechanical test (different method of load introduction) and having absorbed an equivalent amount of moisture may show a substantial loss in the particular

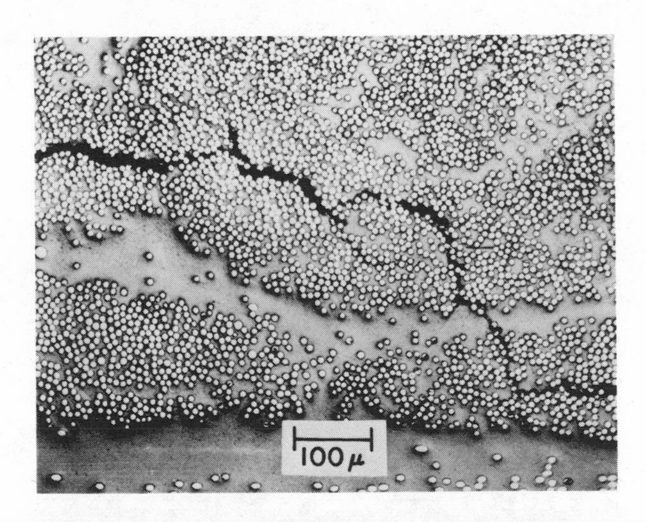

Figure 9. Photomicrograph of humid-aged quasi-isotropic tension specimen (side view, 112.5 ×)

mechanical property value being measured. This behavior is most aptly demonstrated by the boron/epoxy composite system. A quasi-isotropic laminate tested in tension at 350°F after 30 exposure cycles has essentially the same tensile properties as it did at 350°F before any exposure even though it has absorbed a significant amount of moisture. This same boron/epoxy system, having a unidirectional lay-up and being tested in flexure, shows almost a 50% loss in its 350°F flexural strength after it has absorbed an amount of moisture equivalent to that picked up by the quasi-isotropic/tension laminate.

Water behaves as a plasticizing agent, apparently disrupting the strong hydrogen bonding present in the highly polar epoxy systems. Evidence for this are the reversibility of the water absorption effect and the change of failure mode from dry to wet specimens. The test results after water boil exposures can be correlated with those of high humidity exposures based on equivalent water weight gains. The use temperatures of several resin systems are too close to their heat distortion temperatures. The high heat distortion temperature resins are not as significantly affected by moisture at 350°F as are the lower heat distortion temperature resins. The mechanical properties of both "wet" and "dry" composites (all systems) are essentially unaffected up to 250°F.

The effect of moisture is reversible. Drying of wet test specimens results in the recovery of the original dry strengths. The cast resin systems are not hydrolytically unstable as evidenced by the reversibility of moisture absorption.

RECEIVED October 11, 1973.

15

Properties of Filled Polyphenylene Sulfide Compositions

H. WAYNE HILL, JR., ROBERT T. WERKMAN, and G. E. CARROW

Phillips Petroleum Co., Research and Development Dept.,
Bartlesville, Okla. 74004

Blends of polyphenylene sulfide and various fibrous fillers yield a variety of new reinforced thermoplastics which can be readily injection molded. Compounds containing 40% glass fiber combine good processability with excellent mechanical properties. Fybex inorganic titanate fiber and asbestos which can be used at levels up to 20% reinforce to a lesser extent than glass fiber. The glass-reinforced compounds are equal in mechanical properties to other glass-reinforced thermoplastics up to 300°F and generally are superior from 300°–500°F. The compounds are nonburning, possess excellent electrical properties, and have superior resistance to a variety of chemical environments. Compression molded compounds and applications for various filled polyphenylene sulfide compositions are also discussed.

Polyphenylene sulfide (PPS) is a newly available, commercial resin which possesses the unusual combination of high thermal stability and outstanding chemical resistance. The polymer shows an excellent affinity for a variety of reinforcing fillers in both injection and compression molding compositions. These properties and the availability of the resin in several different grades provide utility in various coating applications as well as injection and compression molding markets (*1, 2, 3*).

Ryton PPS grade V-1 is a substantially linear phenylene sulfide polymer of relatively low molecular weight as indicated by a melt flow of greater than 2000 grams/10 min measured in a melt indexer at 315°C with a 5-kg weight. Grade V-1 is a crystalline polymer melting at *ca.*

285°C and is useful for coating applications and as a feedstock for preparing molding compounds (*1, 2, 3*). When grade V-1 polymer is heated in air above about 260°C, chain extension and crosslinking occur, producing a cured polymer of higher molecular weight. Ryton PPS grade P-4 is a cured polymer with a nominal melt flow of 50 grams/10 min and is intended for injection molding alone or in combination with various fillers.

Injection Molding Compounds

Preparation. Any filler which is stable at the 600°–700°F processing temperatures required can be used in PPS molding compounds. Fibrous fillers such as glass are of particular interest since they provide the greatest improvement in mechanical properties at the lowest cost. Injection molding compounds are prepared by dry blending the desired filler with PPS grade P-4. Intensive dry mixers can be used for blending particulate fillers and for some short fiber fillers such as Fybex inorganic titanate fiber when low rotor speeds are used. Cone blenders or drum tumblers are preferred for dry blending chopped glass to minimize fiber damage. Glass fibers can be most readily incorporated into PPS in the form of roving introduced through the vent of a twin screw extruder. The preferred glass is of low (0.3%) sizing level and free of coupling agents since these are prone to decompose and generate gas at the high temperatures involved. Also, PPS bonds well to glass, and coupling agents are not required. The glass fibers in the PPS compounds prepared in a twin screw extruder have an average aspect ratio of 20. The filler content of these compounds is limited by the injection molding process and ranges from about 20% for asbestos or Fybex to about 40 wt % for glass fiber. The injection molding flow decreases sharply as the percentage of filler is increased above the levels indicated for the respective fillers.

Table I. Injection Molding Conditions

Stock temperature, °F	600–700
Mold temperature, °F	150–300
Injection pressure, psi	12,000–18,000
Fill speed	fast
Nozzle	nondrool
Hold pressure, psi	5,000–10,000
Back pressure	none or low
Gate size, inch	0.050–0.30

Molding Conditions. PPS compounds can be injection molded preferably in reciprocating screw machines under conditions similar to other filled resins except for higher stock temperatures. These conditions are listed in Table I. PPS compounds have a slight tendency to gas at molding temperatures. Thermogravimetric analyses show that this amounts to about 0.1%/ hr at 600°F and about 0.5%/hr at 700°F. To prevent void formation especially in thicker parts, the lower stock temperatures along with high injection and hold pressures are recommended. With proper molding conditions and good mold design, well formed parts free of voids and sink marks can be produced. The molding conditions will have some effect on the properties of the molded parts. Higher mold temperatures favor higher tensile properties and better surface appearance with some loss in impact strength. The properties of the compounds reported here are based on specimens molded at 600°F stock temperature, 150°F mold temperature, 15,000 psi injection pressure, and 5,000 psi hold pressure. The compounds have been remolded through three cycles with only 10% loss in tensile strength.

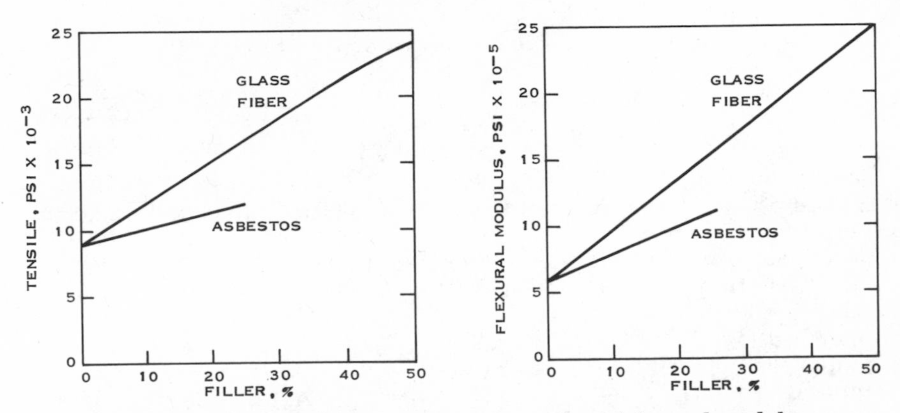

Figure 1. Effect of filler content on tensile and flexural modulus

Properties. As shown in Figure 1, tensile strength and flexural modulus of PPS increase markedly with increasing glass fiber content, and at 50 wt % glass, tensile is increased more than twofold and modulus more than threefold. In comparison asbestos imparts only a moderate increase in these properties up to 25 wt % in PPS. The lower reinforcement realized from asbestos compared with glass is attributed to poorer dispersion and lower adhesion with asbestos. While not shown, Fybex imparts about the same improvement in properties as glass fiber up to 20 wt %. Asbestos and Fybex fillers reduce the melt flow of PPS compounds more drastically than glass fiber. Thus, injection moldable PPS compounds are limited to about 20% asbestos or Fybex *vs.* 40% glass. Typical properties of various injection molded compounds are given in Table II. The addition of 20% asbestos to PPS gives a modest increase in tensile and substantial improvements in flexural modulus and heat deflection temperature. The addition of 40% glass fiber to PPS gives the greatest reinforcement and doubles most mechanical properties. The compound with 20% Fybex has properties in between those of the asbestos and glass-filled compounds but is superior to them in color and surface appearance. PPS is among the least combustible of thermoplastics. For example, the minimum concentration of oxygen to support combustion of PPS is 44%. The addition of glass or asbestos to PPS increases slightly the percentage oxygen required. The electrical properties of several PPS compounds have been described (*4*). Although the addition of fillers degrades the electrical properties—Fybex more so than glass—the compounds retain low dielectric constants and dissipation factors over a wide frequency range. Thus these compounds are suitable for electronic applications at high frequencies where dissipation factor is important, and they are finding use in such applications.

Table II. Typical Properties of Reinforced PPS

	PPS	*80/20[a] PPS/ Asbestos[b]*	*80/20[a] PPS/ Fybex*	*60/40[a] PPS/ Glass[c]*
Density	1.35	1.52	1.53	1.65
Tensile, psi	9,000	12,000	15,000	22,000
Elongation, %	3	2	2	2
Flex. mod., psi × 10^{-3}	600	1,000	1,400	2,200
Flex. strength, psi	20,000	22,000	25,000	35,000
Impact, ft-lb/in.				
notched (¼ × ½)	0.4	0.4	1.0	1.0
unnotched (¼ × ½)	2.0	3.0	3.0	4.0
Hardness, shore D	86	89	88	90
Heat deflection at 264 psi, °F	278	>425	>425	>425
Compressive strength, psi	16,000	—	—	21,000
Mold shrinkage, inch/inch	0.01	—	—	0.002

[a] Parts by weight
[b] Johns Manville 7RF1
[c] Owens Corning 497 X5

The effect of temperature on tensile strength and flexural modulus of 40% glass-reinforced PPS is shown in Figure 2. The curves show similar trends with a marked decrease in both properties between 175° and 275°F as the polymer undergoes a glass transition. Above 275°F the properties decline more gradually up to 500°F. Compared with the unfilled polymer, the PPS containing 40% glass fiber has a marked superiority in properties over the temperature range shown. A comparison of tensile strength *vs.* temperature for several polymers reinforced with 30% glass fiber is listed in Table III. These compounds have similar tensiles at 75°F, and all decrease in tensile with increasing temperature.

The PPS compound has somewhat greater loss in tensile between 150°F and 250°F, but retention of tensile above 300°F is superior. Oven aging tests at 400°F have indicated no significant loss in tensile over 1000 hours.

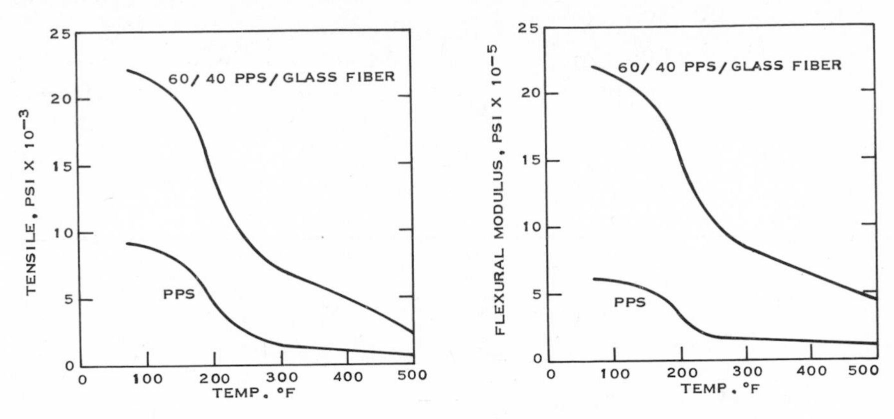

Figure 2. Effect of temperature on tensile and flexural modulus

Flexural creep, which is a measure of the long term load bearing capability of a material, is shown in Figure 3. At 75°F and 5000 psi fiber stress, the glass-reinforced PPS shows low total creep and a very low creep rate through 1000 hours and a very substantial improvement over the base polymer. At 150°F under the same stress, the glass-reinforced PPS shows somewhat higher creep rate up to 500 hours.

Table III. Tensile *vs.* Temperature for Various Glass-Reinforced Polymers

Temp., °F	*Polycarbonate + 30% glass*	*Modified Polyphenylene Oxide + 30% Glass*	*Polysulfone + 30% Glass*	*PPS + 30% Glass*
75	15,600	17,000	15,700	18,000
150	12,000	14,000	13,000	15,000
250	7,000	8,900	11,000	8,000
300	0	5,000	8,000	7,000
350		2,000	4,000	6,000
400		0	0	4,000
500				1,000
550				0

The results of friction and wear tests performed on an LFW-1 friction and wear tester are given in Table IV. Various PPS compounds were tested as flat blocks against standard hardened steel rings. Unfilled PPS showed intermediate friction and wear values. The addition of glass fiber or Fybex gave a small reduction in coefficient of friction but a marked increase in wear. Tests on PPS–fiberglass compounds containing various lubricating fillers showed 5% molybdenum disulfide was ineffective in reducing friction or wear, 9% graphite reduced wear quite markedly, and 25% PTFE reduced both friction and wear. Several of these compounds have friction and wear properties comparable with those of acetal homopolymer and are being further investigated in sliding-friction and journal-bearing tests.

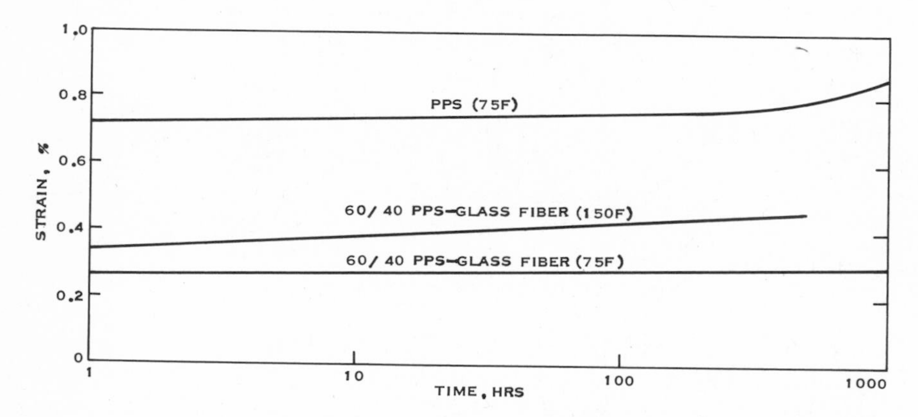

Figure 3. Flexural creep at 5000 psi

Table IV. LFW-1 Friction and Wear Data

Compound	Coefficient of Friction[a] 0 rpm	100 rpm	190 rpm	Wear, mils per 10 min[b]
PPS	0.64	0.64	0.59	8.0
60/40 PPS-glass	0.50	0.55	0.53	>15
80/20 PPS-Fybex	0.52	0.59	0.50	>15
70/25/5 PPS-glass-MoS_2	0.50	0.56	0.53	>15
55/36/9 PPS-glass-graphite	0.50	0.48	0.52	3.7
50/25/25 PPS-glass-PTFE	0.30	0.32	0.35	3.3
75/25 PPS-PTFE	0.27	0.31	0.33	5.2
Acetal homopolymer	0.29	0.32	0.35	2.6
PTFE	0.20	0.26	0.27	1.5

[a] Flat test block, 15-lb load.
[b] Flat test block, 90-lb load.

PSS is resistant to many organic solvents. Tensile specimens exposed at 200°F for 24 hours were unaffected by hydrocarbons, alcohols, ketones, esters, ethers, and organic acids. Under these conditions some chlorinated solvents and nitrogen compounds slowly attack the polymer while others do not. Concentrated inorganic acids (96% sulfuric) and strong oxidizing agents (chromic acid and sodium hypochlorite) attack PPS while weaker acids (30% sulfuric) and strong bases do not. When compared with other polymers exposed under the same conditions (Table V), PPS is resistant to a greater variety of chemical environments than most thermoplastics.

Table V. Chemical Resistance (200°F, 24 hours)

	Tensile Retained, %			
Reagent	*Polycarbonate*	*Modified Phenylene Oxide*	*Polysulfone*	*PPS*
30% H_2SO_4	101	102	105	101
30% NaOH	0	102	101	114
Acetic anhydride	0	55	0	100
Ethylene chloride	0	0	0	72
Toluene	0	0	0	98
Amyl alcohol	48	62	48	102
Methyl ethyl ketone	0	0	0	112
Pyridine	0	0	0	93

Compression Molding Compounds

Preparation. For compression molding, as with injection molding, various thermally stable materials have been used for fillers. In most cases compression molding compounds contain higher loading of filler than the injection molding compounds. Compression molding compounds are prepared by blending the uncured polymer with the desired filler. The uncured polymer has a low melt viscosity and tends to wet the filler better than the more viscous cured material used in the injection molding blends. Standard mixing equipment such as Henschel mixer, drum tumblers, Waring blender, and other intensive mixers can be used for most fillers. The most important considerations in blending are that the filler particles are uniformly distributed throughout the mixture and thoroughly coated with the PPS powder. For glass fibers longer than 1/32 inch, special equipment and techniques are required for good blends. Glass with longer fibers is difficult to separate properly and to obtain thorough coating of the individual fibers.

Both filled and unfilled PPS must be highly cured to reduce the melt flow prior to its use as a compression molding compound. Curing can be done by exposing the mixture to air at elevated temperatures. The time and temperature used are controlled by the bulk density of the mixture. Compounds of low bulk density such as those of high asbestos content can be cured sufficiently in 1 hr at 700°F in an air-circulating oven. Compounds containing short glass fiber and of intermediate bulk density require several hours cure at 550°F and then 1 hr cure at 700°F. Unfilled PPS and blends with powders or pigments are usually cured 16 hours at 510°–540°F and then 1 hr at 700°F. The cure below the melt point of high bulk density blends is required to increase the melt viscosity. This prevents puddling and allows air to contact the PPS throughout the entire mixture during the 700°F cure.

Table VI. Properties of Compression Molded PPS Compounds

	60/40[a] PPS/ Asbestos	*60/40[a] PPS/ Glass*	*60/40[a] PPS/ Fybex*	*PPS*
Tensile, psi				
75°F	10,700	11,000	8,400	4,800
175°F	8,300	8,000	6,600	4,200
275°F	3,200	4,000	4,000	600
400°F	2,200	2,400	2,200	—
Flexural modulus, psi × 10^{-3}				
75°F	1,200	1,200	1,200	420
175°F	1,200	1,000	1,100	370
275°F	190	340	240	10
400°F	100	180	120	—
Flexural strength, psi				
75°F	13,680	11,190	13,790	9,000
175°F	10,420	8,940	12,240	8,600
275°F	6,110	3,520	4,780	—
400°F	3,070	—	3,560	—
Compression strength, psi				
75°F	21,700	21,000	20,500	14,900
175°F	18,800	19,000	17,200	10,600
275°F	16,300	15,000	13,800	[b]
400°F	8,800	9,000	7,100	[b]
Thermal expansion, inch/inch/°F × 10^{-5}				
75°–200°F	2.0	2.0	2.0	—
200°–300°F	3.5	4.2	3.0	—
300°–400°F	5.5	6.9	6.1	—

[a] Parts by weight.
[b] Ductile, showed no yield point or break.

Curing is usually done by placing the blended material in an open topped pan at depths of ½ to 2 inches. Compounds of low bulk density can be cured with bed depths up to 2 inches while powder blends and unfilled PPS of high bulk density should be cured at approximately ½ inch. The cured material usually forms a sheet or solid mass slightly more compact than the uncured material. This mass requires granulating to particles approximately ⅛ inch in diameter before molding.

Molding Procedure. Molds which can withstand pressures of 4000 psi at 750°F are required for compression molding PPS compounds. In most cases steel molds have been used. To prevent sticking and to improve the part finish, the mold should be clean and coated with a mold release. The prepared mold is filled with molding compound and cold pressed at 2000–3000 psi pressure. The filled mold is then heated until a minimum temperature of 600°F in the mold is reached. This usually requires from 1 to 3 hours in a 650°–750°F oven. The hot mold is then placed in a press and subjected to 1000 to 4000 psi pressure, thicker parts requiring the higher pressures. Controlled cooling is required to prevent voids or cracking of large parts. A maximum cooling rate of approximately 4°F/min must be held until the temperature is below 450°F. The part can be removed from the mold after the temperature reaches 300°F. This technique has been used to mold rods, tubes, slabs, and preforms which only required slight machining for finished, close-tolerance parts.

Properties. The compression molded compounds containing 40% asbestos, glass fiber or Fybex display a substantial increase in mechanical properties over the unfilled resin as shown in Table VI. The properties of the filled compounds are similar over the temperature range shown, and consequently the fillers appear equal with respect to properties imparted to the compounds. The properties of the compression molded compounds are well below those of the 40% glass reinforced injection molding compound. The difference in properties is attributed to the low crystallinity of the highly cured compression molding resin as indicated by the low heat of fusion by DTA. Linear coefficient of expansion of these compounds is low and constant from 75°F to 200°F, the glass transition. Above 200°F the coefficient of expansion gradually increases with increasing temperature.

Applications

In summary, polyphenylene sulfide compounds offer a combination of desirable properties such as good thermal stability, outstanding chemical resistance, low coefficient of friction, excellent electrical properties, and precision moldability. These properties lead to applications not available to many other plastic materials. For example, a number of pump manufacturers are using PPS compounds as sliding vanes, impeller cases, gage guards, and seals in corrosive service involving materials such as 60% sulfuric acid, liquid ammonia, and various hydrocarbon streams. Polyphenylene sulfide compounds with low friction and low wear properties have been evaluated as cages for non-lubricated ball bearings at

350°F and 50 psi load. These materials have operated longer than 600 hrs while other materials fail in less than 20 hrs. In another type of application, electrical properties and the ability to injection mold very small parts with great precision have led to the use of a variety of connectors, coil forms, etc., in the electronics industry. In another application a 10.5-inch diameter piston for a non-lubricated gas compressor has been in service at 1000 rpm for over 6 months and is performing better than the aluminum piston it replaced. This piston was machined from a 35-lb compression molded block of PPS. In addition to these uses for various molded parts, filled polyphenylene sulfide coatings are finding utility as release coatings for cookware and as corrosion resistant coatings for the chemical and petroleum industries.

Literature Cited

1. Hill, H. Wayne, Jr., Edmonds, J. T., Jr., *Preprints, Amer. Chem. Soc., Div. Org. Coatings Plastics Chem.* (1970) **30** (2) 199.
2. Short, J. N., Hill, H. Wayne, Jr., *ChemTech* (1972) **2**, 481.
3. Hill, H. Wayne, Jr., Edmonds, J. T., Jr., *Preprints, Amer. Chem. Soc., Div. Polym. Chem.* (1972) **13** (1) 603.
4. Hill, H. Wayne, Jr. *et al.*, ADVAN. CHEM. SER. (1973) **129**, 80.

RECEIVED October 11, 1973.

16

Biodegradable Fillers in Thermoplastics

GERALD J. L. GRIFFIN

Brunel University, Department of Polymer Science and Technology, Uxbridge, Middlesex, England

Increased degradability in landfill and composting of the common packaging thermoplastics has been achieved by incorporating a biodegradable filler into the plastics compounds using standard hot-melt compounding techniques. A search of possible fillers disclosed that only raw starch satisfied the requirements of adequate thermal stability, minimum interference with melt-flow properties, and minimum disturbance of product quality. Successful extrusion-blowing of layflat film in LDPE containing up to at least 30 wt % of starch is reported, and starches—principally maize, rice, and tapioca—have been successfully included in other products such as fibrillated PP film, TPS injection moldings, extrusions, and thermoformings. The characterization of starches by scanning electron microscopy and narrow angle light scattering is described as part of this investigation.

The preservation of our environment has excited much public and technical discussion in recent years. Special attention has been focused, perhaps somewhat unfairly, on the particular problems associated with the increasing proportion of plastics packaging materials in community domestic refuse. Wallhäuser has published three detailed articles (*1, 2, 3*), which describe the situation in the Federal German Republic but nevertheless are of general interest especially in connection with composting and landfill procedures. Polyolefin film has received special criticism because of its longevity under soil burial conditions. Because the only likely degradation processes acting on buried polyolefins are simple oxidation and microbiological attack, the valuable work of Scott (*4*) on the controlled photooxidation of polymers cannot provide a solution. Wallhäuser reports that in five-year soil burial tests LDPE had been colonized by *Desulfovibrio* bacteria, and material from 2.5-meters depth had become embrittled. More recently Nykvist (5) has established that LDPE is slowly biodegraded in compost; he demonstrated this by experiments using carbon-14-labeled polymer and detecting radioactive CO_2 in the air aspirated from the sample vessel. Long term oxidation studies of LDPE, which are of great interest to cable and pipe manufacturers, have been well documented, and power factor measurement is the preferred method for monitoring the oxidation. The early work is reviewed by Haywood (6) and suggests that the oxidation rates are negligible at room temperature, but 1-mm thick sheet would embrittle in about one year at 40°C. However, temperatures above 40°C are common in composting. Also, oxygen diffusion rates in the polymer give extra significance to film thickness and temperature.

Evidently one circumstance which would encourage both modes of attack would be an increase in the specific surface area of the material; Wallhäuser refers to the desirability of shredding the waste before burial. An alternative approach is to introduce a filler into the material which itself is speedily degraded, thus leaving a porous film readily entered by microorganisms and rapidly saturated with oxygen. The selection of a suitable filler is the subject of this work.

Criteria for Biodegradable Filler Selection

The markets currently satisfied by plastics packaging films expect the industry to supply them with a product that is strong, smooth, odorless,

Figure 1. Scanning electron microscope picture of rice starch grains. Average grain size is about 5 μm.

basically colorless, non-toxic, water resistant, and cheap. Any biodegradable filler must not unduly compromise these qualities and must be able to withstand processing temperatures for short periods. These normally fall within the range 150°–300°C. The primary requirement of biodegradability is achieved by the filler acting as a potential major nutrient for some microorganisms or being decomposed by microorganism nutrition. This means, inevitably, that the filler will be organic in nature. Low cost requirements suggest that a waste product from another industry should be considered, but an interesting survey of industrial wastes by Gutt (7) reminds us that the overwhelming bulk of industrial waste is mineral in origin. Certain well known low cost organic wastes such as lignin and leather grindery refuse are immediately eliminated because of color or odor, and protein-based materials are too thermally unstable to be considered. Cellulosic materials are traditional ingredients of thermosetting molding compositions, and World War II experience with military equipment in the tropics established the accessibility of these fillers to biological attack. The physical nature of wood flour and cellulose pulp make them unwelcome high-volume additives in thermoplastics because

Figure 2. Scanning electron microscope picture of potato starch grains. Average grain size is about 50 μm.

the refined nature of the extrusion-blowing and extrusion-coating technologies make them very sensitive to changes in their rheological properties.

Minimum particle/particle interaction in flowing suspensions, and hence minimum viscosity increase, is achieved by systems in which the suspended particles are smooth spheres or, even better, smooth ellipsoids which can orient to a minimum energy configuration in laminar flow.

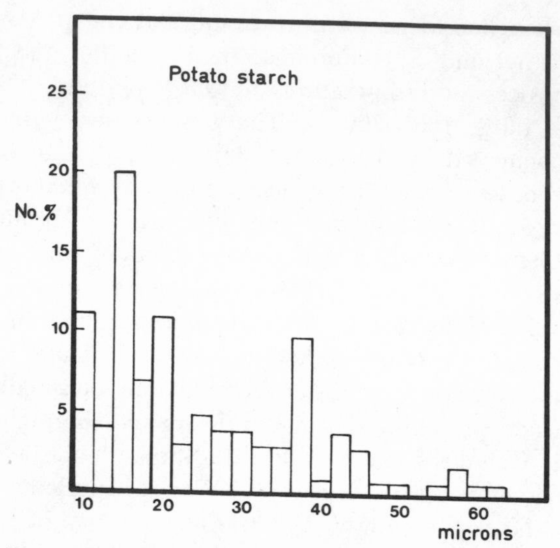

Figure 3. Potato starch mean diameter histogram

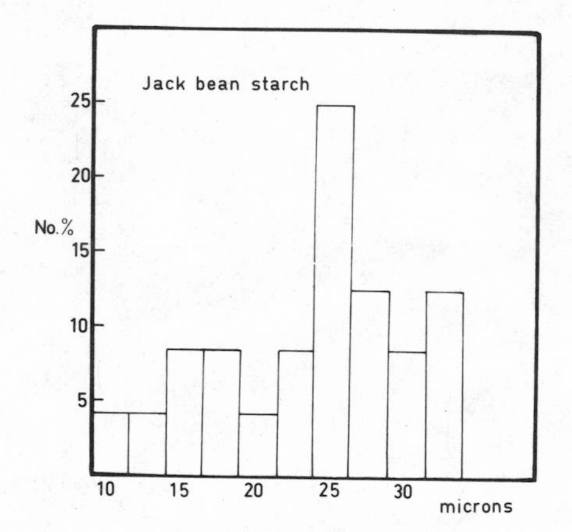

Figure 4. Jack bean starch mean diameter histogram

This theory has been long established for Newtonian fluids and is reviewed and extended in the second Dutch report on viscosity and plasticity, especially by Burgers (*8*). A further rheological consideration is the question of particle size distribution. Eveson (*9*) found reductions of up to 16% in the relative viscosity of 22.5% microsphere suspensions by changing only the size distribution away from the homodisperse.

These considerations narrow the field of search to particles of regular geometries, and only spores, seeds, dried simple organisms, and starches remain for consideration. Starches are the most attractive members of this group and were selected for further study (*10*).

Characteristics of Starches

Little needs to be recorded here on general starch technology other than a reference to the remarkable 'atlas' of Reichert (*11*) and the recent comprehensive text of Whistler and Paschall (*12*). The criteria set out earlier are met to a remarkable degree by a small group of commercially available starches. The particles can be near spheres, as in the polyhedral rice starch grains shown in Figure 1, or elongated near ellipsoids as with the potato starch grains shown in Figure 2. These geometrical variations are determined by the mode of occurrence of the starch grains within the parent plant structures, compound starch grains giving rise to the faceted particles. The range of particle sizes is from 3 to 100 μm, and their

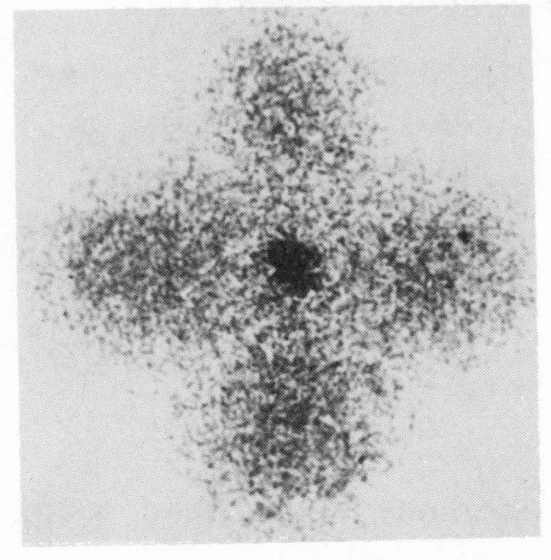

Figure 5. Narrow angle light scattering pattern directly recorded on photographic plate following method of Stein and Rhodes. Sample was dispersion of wheat starch in Canada balsam solution in xylene held as film between microscope slide and cover slip.

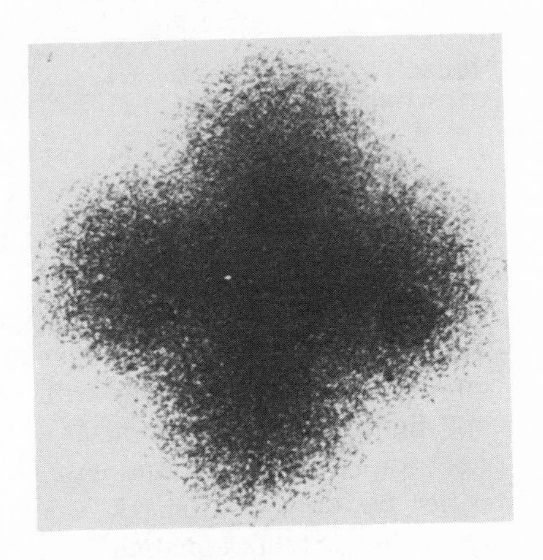

Figure 6. Same as Figure 5, but sample was dispersion of maize starch

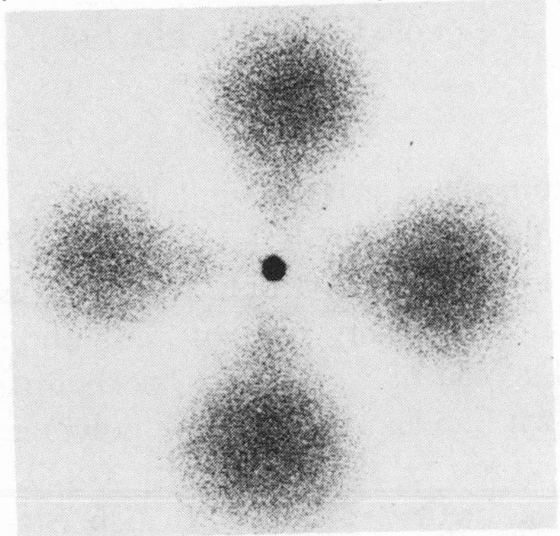

Figure 7. Same as Figure 5, but sample was dispersion of rice starch

measurement was considered important in this work. Stein and Rhodes's method of spherulite size measurement (*3*) is directly applicable to starch suspensions, and Samuels (*14*) has used rice starch as calibrant particles in light scattering work. Some caution is indicated because of the occurrence of skewed and double-peaked populations. Sample histograms from direct microscopy are shown in Figures 3 and 4. The quality of the H_v scatter cloverleaf varies greatly between starches, and a series of examples progressing in mean particle diameter are shown in Figures 5, 6, and 7. There is also occasional uncertainty about the starch grain dimensions quoted in the literature because the longest chord measurement on strewn samples can be quoted without qualification. Table I compares my measurements with certain published values. The microscopy figures are of the form $D = (l + b)/2$ and were determined on photographs of settled suspensions. The narrow angle light scattering results are derived from H_v mode He-Ne laser light scattering experiments using Stein's equation

Table I. Mean Diameters of the Common Starch Grains[a]

Starch	*Microscopy*	*Light Scattering*	*Literature*
Rice	5	5.6	3 to 8
Maize	11	14.2	Av. 15
Arrowroot	21	25.8	—
Wheat	26	36.4	2 to 10 and 20 to 35
Jack bean	24.5	30.8	—
Potato	25.5	55.2	15 to 100

[a] All dimensions in micrometers.

$$D = 2\lambda/\pi \sin (\theta/2)$$

The literature figures are quoted from Knight (*15*).

Compounding

Acknowledged difficulties in relating laboratory rheometry to polymer processing technology encouraged me to adopt a "titration" technique for assessing the compatibility of starch with LDPE. A set weight of polymer was fluxed on a laboratory two-roll mill, and a starch was added progressively until the hide of compound broke up or became unmanageable. For the blown film experiments the polymer used was Imperial Chemical Industries grade Q1388 of density 0.920 at 23°C and melt flow index 2 following method 105C of British Standard Specification 2782. Rice, wheat, potato, maize, and tapioca starches were all blended easily into the fluxed polymer provided only that they were dry and not in hard agglomerates. Milling times exceeding 1 hr at roll temperatures of 150°C caused no visible increase in discoloration, and subsequent light microscopy of films made from these materials showed discrete intact starch particles in the polymer matrix. Controlling size distribution by using blends of small particle and large particle starches enabled weight loadings exceeding 100 phr to be achieved. These concentrated stocks were granulated for use as masterbatches by cutting the stock as strip from the mill and feeding, after air cooling, to a rotary cutter.

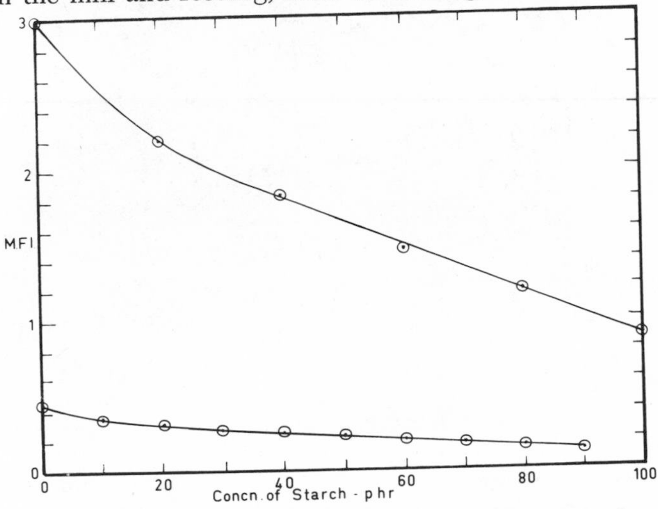

Figure 8. Melt flow index for two grades of LDPE with increasing starch content. Maize starch only was used.

Conversion Processing

Initial work was restricted to sample sheet preparation by compression molding, heat stability testing by the extrusion of a 12 × 1-mm ribbon, and extrusion-blown film trials using a line based on a Samafor extruder of 20:1 *l:d* ratio with a 45-mm diameter screw making 300-mm wide layflat at 25 micrometer nominal gage.

The heat stability procedure involved continuous extrusion at the lowest machine speed of an LDPE ribbon containing 10 phr starch while progressively raising the head and die temperatures. Taking maize starch as typical, some vapor evolution occurred at 230°C which was presumably caused by steam generation, and it caused a roughening of the surface of the extrudate. The temperature limit of the equipment was 300°C, and at this level a pale cream color developed in the products. The onset of discoloration was at a higher temperature than expected, and it would seem that the unusual environment of the isolated starch grains may be a significant retarding factor.

Film extrusion-blowing was evaluated by setting the Samafor film extrusion line to operate normally with unmodified LDPE and then progressively increasing the content of 100 phr starch masterbatch in the feed blend. Provided only that the masterbatch was dry, the starch/ LDPE compound extruded as readily as the unmodified material with

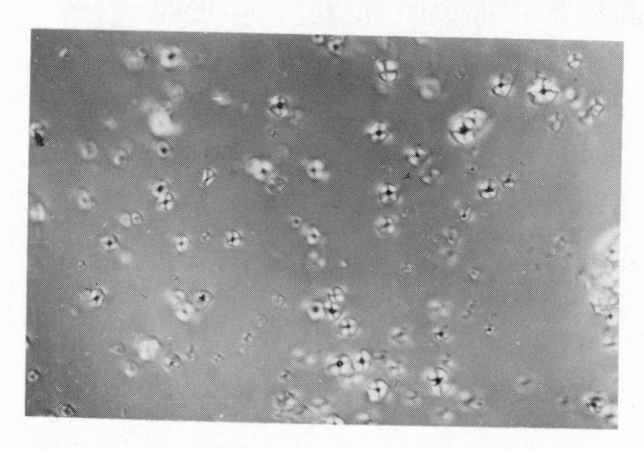

Figure 9. Rice starch grains, 5 μm mean diameter, seen by polarized light microscopy in polypropylene film after extrusion at 230°C

no indication of increased power consumption on the normal machine instruments. Above 30 phr occasional bubble failures occurred, but it was possible to blow the 100-phr masterbatch directly for short periods. Early difficulties with screen pack blockage were later avoided by careful attention to the control of masterbatch compounding technique. Melt flow index measurements on a series of LDPE/maize compounds were made and confirmed the modest interference with flow properties suggested by the extrusion trials; the results are presented in Figure 8. The LDPE film produced was similar in feel and appearance to unmodified film when the starch content was low, apart from an increase in surface roughness which eliminated the blocking tendency of the film and decreased its transparency. At starch concentrations above 15 phr the product began to develop a pleasing, papery feel which was very pronounced at 30 phr.

In view of the success of these trials the work was extended to establish the compatibility of starches with other thermoplastics by small scale mill compounding. So far, this has been verified for all the common packaging thermoplastics. With polypropylene it was easier to extrude a blend of PP granules and LDPE/starch masterbatch granules, a technique adopted to avoid exposing starch to high temperatures while unprotected by a polymer melt envelope. Even after extrusion in PP at 230°C the starch grains were undamaged as indicated by their appearance under polarized light microscopy (Figure 9). Extruded PP ribbon with 15-phr rice starch content could be cold-drawn at up to 12:1 ratio for fibrillation.

The polystyrene compounding trials were scaled up first to a Shaw 3-kilo internal mixer, and later to a Buss PR100 continuous mixer equipped with a face cutter for pellet production. The product was used for making extruded sheet from blends with toughened polystyrene, using a vented barrel Samafor extruder with a 45-mm diameter screw of 25:1 *l:d* ratio and coupled to a 300-mm slit die. The sheet performed well in thermoforming trials of deep disposable drinking cups. The same polystyrene master batch was used in blends for injection molding trials and, once again provided only that the masterbatch was dry, no difficulties were experienced with molding the customary test pieces.

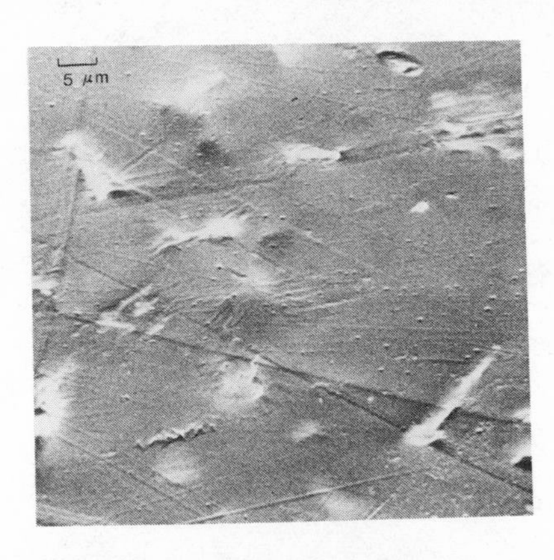

Figure 10. Scanning electron microscope picture of 30 phr maize starch-filled LDPE sheet, pressed sample, untreated

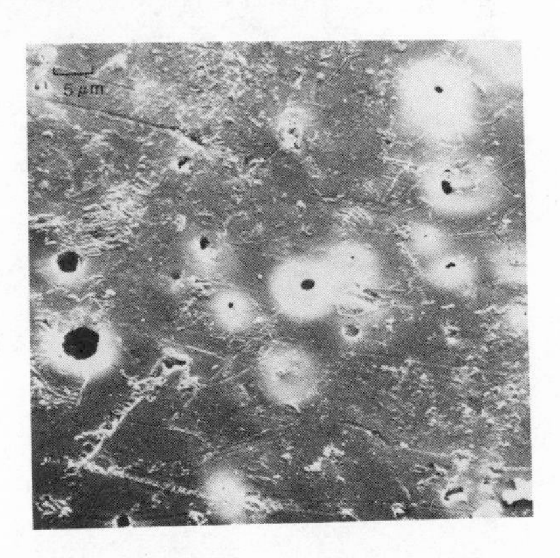

Figure 11. Scanning electron microscope picture of same sample as in Figure 10 after 24 hr enzyme extraction at 35°C

Biological Testing

The time problem associated with soil burial tests was circumvented to a degree by adopting a direct enzyme attack procedure. The presence of amylase sources in most soils is assured by the wide distribution of organisms such as *B. subtilis*. For convenience an α-amylase concentrate, Sigma Chemical Co. A-6755, from malt was used in the form of a 0.1% solution in water with appropriate salts and buffers and held at 35°C in an incubator. Polyethylene/starch sheet samples were suspended in this enzyme solution for 1 to 10 days when, typically, the surfaces would become slimy probably because of a layer of limit dextrans which were rinsed off before the samples were dried at room temperature, vacuum metallized with Au/Pd, and examined by scanning electron microscopy. Figures 10 and 11 show a pressed LDPE sheet containing 30 phr maize starch before and after enzyme treatment. Control samples incubated in water over the same period showed no apparent change. The small surface disturbances created by the underlying starch grains in the untreated sheet have developed, after enzyme exposure, into deep pits; these proved to be rather difficult objects for scanning electron microscopy because of the failure of the metallization to penetrate into the undercuts, giving rise to the familiar electrostatic "flare."

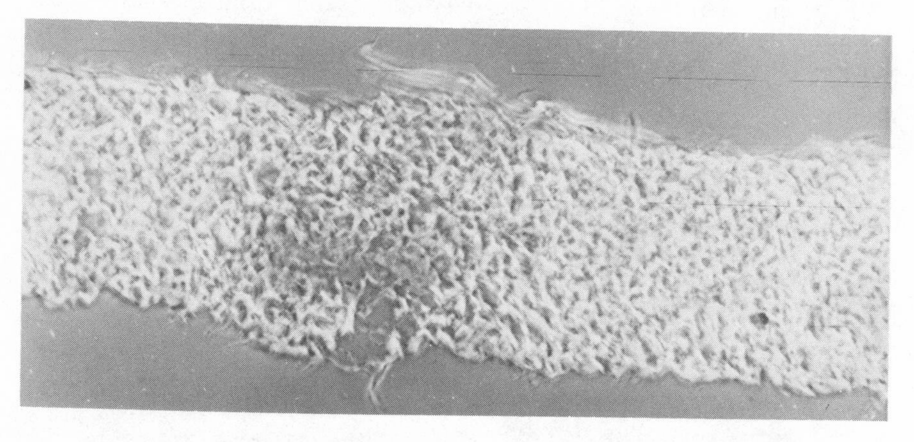

Figure 12. Microtome cross section through 50-μm thick LDPE blown film originally containing 30 phr starch but extracted for 10 days with enzyme solution at 35°C. Phase contrast transmitted light photography.

Further experiments were conducted in order to show the extraction effect in depth. For example, extrusion-blown LDPE film containing 30 phr maize starch was extracted as above for 10 days and then embedded in water-soluble wax for microtomy. Figure 12 shows the appearance of such a sample in cross section as seen by phase contrast microscopy, and it is evident that the material has been converted into a sponge. In another material, thin sheets of polystyrene containing 25% tapioca starch were enzyme-extracted and the change was revealed by staining the starch grains with iodine solution. Figures 13 and 14 show this 125-μm thick polystyrene treated identically except for the enzyme extraction. This work has been paralleled by soil burial experiments which indicate that similar events take place. For example, by following weight loss it has been established that 80% of the starch in an LDPE 50-μm thick film containing 15 phr maize starch has been extracted in moist garden soil in about 8 weeks at 25°C (*16*).

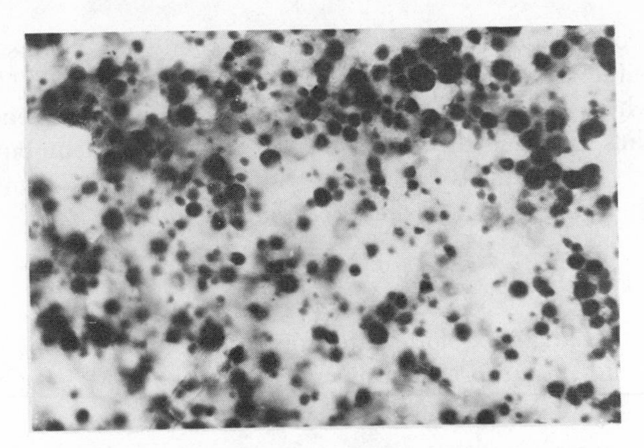

Figure 13. Toughened polystyrene sheet containing 25 phr tapioca starch photographed by transmitted light microscopy after staining starch with iodine

Figure 14. Piece from same sample sheet used for Figure 13 but stained and photographed after 10 days enzyme extraction at 35°C. Holes left by absent starch grains can just be seen.

Conclusions

A biologically innocuous filler has been selected which appears to cause minimum disturbance to the common packaging thermoplastics as far as their processing and properties are concerned at loadings of up to 10 phr; above this level it produces compositions which are attractive in their own right. In LDPE films these more heavily filled compositions have a papery quality. This filler, or rather group of fillers—the plant starches, resists dry heat in the normal plastics processing operations. They are inert to water at room temperature but are rapidly digested by the ubiquitous amylases; the porosity thus generated offers the most favorable circumstances for biological and oxidative attack. The mechanism by which the large enzyme molecules are able to penetrate the polymer films between the starch particles is not evident but is being investigated. Trials are also being extended to long term soil and compost burials and to other polymers and processing methods.

Acknowledgments

The author thanks Coloroll Ltd. for their support of this work, and final year students at Brunel University, especially Mr. Peter Jenkins, for patient manipulation of microscope and laser.

Literature Cited

1. Wallhäuser, K. H., *Verpackungs Rundschau* (1972) **3**, 266.
2. Wallhäuser, K. H., *Mull Abfall* (1972) **1**, 10.
3. Wallhäuser, K. H., *Preprints,* Degradability of polymers and plastics, Plastics Institute, London, 27 Nov. 1973.
4. Scott, G., *Plastics Rubbers Textiles* (1970) **1**, 361.
5. Nykvist, N., *Preprints,* Degradability of polymers and plastics, Plastics Institute, London, 27 Nov. 1973.
6. Haywood, C. K., in "Polythene," Renfrew and Morgan, Eds., p. 135, Iliffe, London, 1960.
7. Gutt, W., *Chem. Ind.* (1972) 439.
8. Burgers, J. M., 2nd, Report on Viscosity and Plasticity. Verhandelingen der Konok. Nederl Akad. Eerste Sektie Deel XVL No. 4 North Holland Pub. Co. Amsterdam 1938.
9. Eveson, G. F., in "Rheology of Disperse Systems," pp. 61-83, Pergamon Press, London, 1959.
10. British Patent Application **23469/72**, Assigned to Coloroll Ltd.
11. Reichert, E. T., *Carnegie Inst. Washington Pub. No.* **173** (1913).
12. Whistler, R. L., Paschall, E. F., "Starch Chemistry and Technology," Academic Press, New York, 1965.
13. Stein, R. S., Rhodes, M. B., *J. Appl. Phys.* (1960) **31**, 1873.
14. Samuels, R. J., *J. Poly. Sci.* (1971) **A2, 9**, 2165.
15. Knight, J. W., "The Starch Industry," Pergamon Press, London, 1969.
16. Dowding, P., private communication, Trinity College, Dublin, 1974.

RECEIVED October 11, 1973.

17

Performance of Conductive Carbon Blacks in a Typical Plastics System

J. H. SMUCKLER

Conductive Polymer Corp., Marblehead, Mass. 01945

P. M. FINNERTY

Cabot Corp., Billerica, Mass. 01821

Carbon black, used primarily as a reinforcing filler in rubber, exhibits properties of electrical conductivity and physical form which render it the preferred filler for imparting conductivity to normally insulative elastomeric and thermoplastic polymers. The properties of carbon black important in imparting conductivity to polymer systems are discussed. The effects of three different carbon black types at varying concentrations on the electrical, physical, and rheological properties of poly(ethylene vinyl acetate) (EVA) in both thermoplastic and thermoset form are also illustrated. Reference is also made to the phenomenon of significant resistivity variations with increasing temperature of carbon black-loaded plastics systems.

Conductive carbon black is the preferred filler for imparting anti-static and conductive properties to polymer systems. Conductive plastics compounds containing carbon black are widely used in the wire and cable industry as metal conductor strand and primary insulator shielding in high voltage cables. Similar plastics compounds are used to produce anti-static sheeting, belting, hose, and molded goods which minimize static buildup in hazardous environments such as mines and other areas where explosive vapors may collect.

This paper reviews briefly the reasons carbon black is classified electrically as a "semi-conductive" material and how these conductive properties are imparted to plastic systems. Primarily, however, we illustrate the effects of three carbon black types at various concentrations on the electrical, physical, and rheological properties of ethylene–vinyl acetate copolymer, in both thermoplastic and thermoset (chemically crosslinked) form.

Carbon Black Properties Which Affect Conductivity

Figure 1 shows the relationship between the conductivity of carbon black, pure polymers, carbon black-filled polymers, and metals. Semi-conductive materials, as opposed to metals, exhibit a negative temperature coefficient of resistance, showing an increase in conductivity with increasing temperature. Furnace blacks, such as the grades discussed here, have very small negative temperature coefficients and thus vary only slightly in dry resistivity with temperature variation.

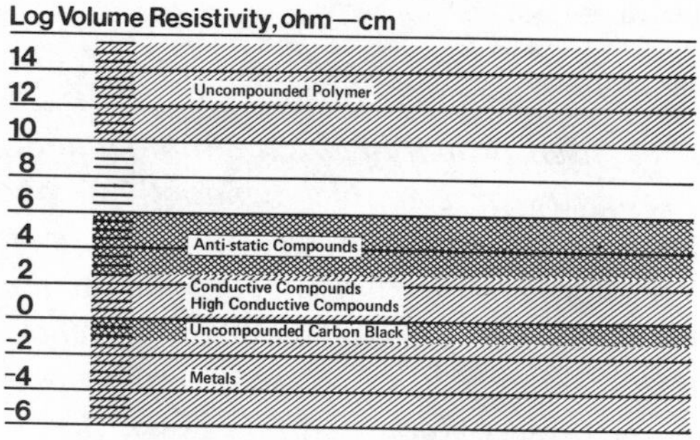

Figure 1. Volume resistivity of polymer compounds and reference materials

These carbon blacks of low volatile content (chemisorbed oxygen complexes) exhibit dry resistivities in the range of 0.2 to 1.0 ohm-cm measured at a bulk density of $\sim$0.5 gram/cm^3 at 25°C. As discussed later, certain carbon blacks containing large amounts of these oxygen groups on their surfaces can exhibit resistivities as high as 50 ohm-cm (*1*). Carbon black is routinely used to impart electrical conductivity to various insulative polymeric systems including the more common thermoplastics because of its finely divided powdered form. Although all carbon blacks are semi-conductive, certain grades are preferred for compounding into polymer systems. The reasons for this are discussed below. Further information on the more theoretical aspects of conductive carbon blacks in plastics systems is available in some recent work of Mildner (*2*), and Forster (*3*).

Carbon Black Properties Which Affect Conductivity of Plastics Systems

Carbon black is composed of particles fused together to form nodular aggregates. In most cases, these aggregates do not break up during dispersion and processing. Rather, the dispersion process consists of breaking up large agglomerates (consisting of a great number of these aggregates compacted together). Dispersion is considered complete when most of these aggregates have been separated and surrounded by polymer.

For the reasons given above, carbon black is a conductor compared with most polymers (*see* Figure 1). Several theories have been proposed to explain how carbon black imparts its electrical conductivity to plastics products, but the full scope of these theories is beyond this discussion. In general, however, for current to flow through a conductive plastics compound, electrons must travel *via* the carbon black since the polymer phase is an insulator (*4*). To achieve this continuous electron flow, the carbon black aggregates must be in contact or very close together. The greater the number of aggregates in contact or close together, the greater the conductivity of the resulting compound (*4*). Therefore, those factors which increase aggregate-to-aggregate contact or reduce the distances between aggregates tend to increase conductivity. Researchers have estimated this distance to be someyhat less than 100 A for electrons to flow across the insulative polymer barrier (*1, 4*). It is felt electrons

Figure 2. Effect of nodule size on conductivity

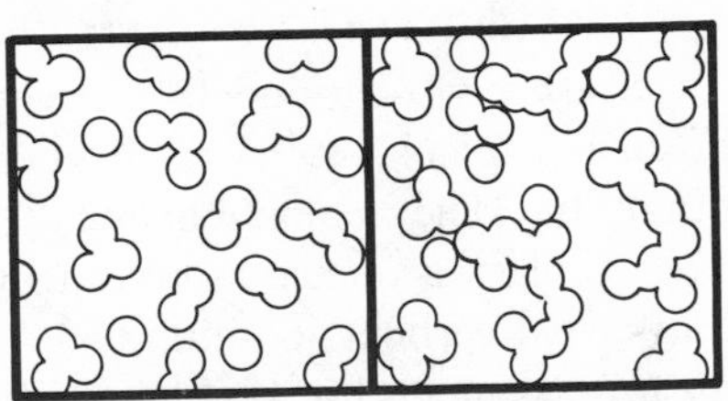

Figure 3. Effect of aggregate shape (structure) on conductivity

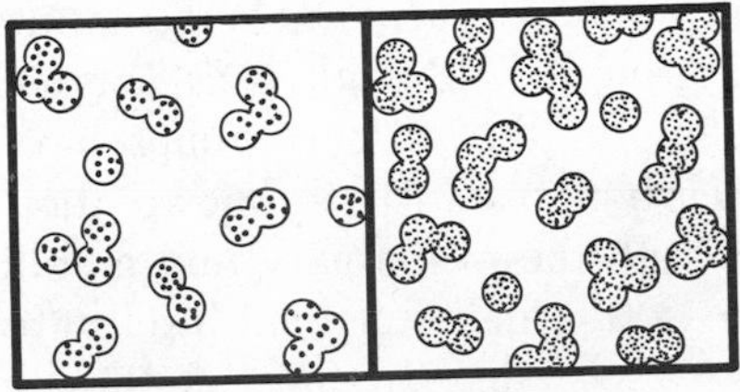

Figure 4. Effect of porosity on conductivity

penetrate this barrier *via* the mechanism known in the semi-conductor industry as electron "tunneling" (*1, 5*).

Three basic carbon black properties affect the interaggregate distance for a given carbon black loading in a polymer system: nodule or primary particle size, aggregate shape or structure, and porosity (*1, 4*). Figure 2 illustrates the effect of nodule size on conductivity. Since compounding ingredients are added on a weight basis, the addition of small nodule-size carbon blacks would, in effect, add more aggregates to the mix than a larger nodule-size carbon black. As such, smaller nodule- or primary particle-size carbon blacks would have more aggregates in contact as well as separated by small distances, resulting in greater conductivity.

The effect of carbon black structure on conductivity is illustrated in Figure 3. High structure carbon blacks tend to produce larger numbers of aggregates in contact as well as separated by small distances, resulting in higher conductivity.

Figure 4 shows the effect of carbon black porosity on conductivity. Since on a weight basis the more porous carbon blacks are, in effect, resulting in more aggregates being added to the compound, the interaggregate distance is also decreased, resulting again in higher conductivity.

Earlier we alluded to the volatile or oxygen content on the surface of carbon black—its surface chemistry—in terms of the effect on conductivity. The major chemical constituents found on the surface of carbon black are carbon-bound hydrogen, hydroquinones, quinones, lactones, and carboxylic acids (*6*). Although the effect of these functional groups is minor for the low volatile content carbon blacks considered here, a discussion of the factors influencing the conductivity of carbon black would not be complete without some treatment of the effect of surface chemistry on conductivity. Carbon blacks such as the channel process types which contain relatively large amounts (between 5 and 15%) of these groups on their surface are much more resistive than the furnace blacks which contain much smaller amounts ($<2\%$) of these chemisorbed oxygen groups. Although it is not completely clear why some of the constituents of the surface groups have such an effect on the conductivity of carbon black, it is known that these chemisorbed oxygen groups affect both the availability and mobility of the free electrons (*1*). For this reason, usually channel blacks are not used to impart conductivity to plastics systems. In fact, in those plastics pigmenting applications where black color is required and the maintenance of low dielectric constant properties is important, channel blacks at concentrations of 1% or less are preferred because of the electron dampening (*1*) effect of surface chemistry.

Analytical Properties of Conductive Carbon Blacks

Three grades of conductive carbon black, N-472, N-294, and N-293, were evaluated. Table I lists their typical analytical properties. Nitrogen surface area is a measurement of particle (nodule) size since small particles have higher surface area than larger particles. DBP (dibutyl phthalate) absorption is a measure of structure or bulkiness in the aggregates since, all other factors being equal, high structure carbon black absorbs more DBP than lower structure carbon blacks. Based on the previously discussed properties which affect conductivity in plastics, N-472 would be considered the most conductive of the three carbon blacks, followed by N-294 and N-293 in that order.

Table I. Typical Analytical Properties of Conductive Carbon Blacks

ASTM No.[a]	*N_2 Surface Area, m^2/gram*	*DBP Absorption, cc/100 grams*
N 472 (ECF)	254	178
N 294 (SCF)	203	106
N 293 (CF)	145	100

[a] The commercial carbon black grades used in this work were:
N 472—Cabot Corp. Vulcan XC-72;
N 294—Cabot Corp. Vulcan SC;
N 293—Cabot Corp. Vulcan C.

Table I shows why this is so. Grade N-472 is highest in surface area, meaning smaller nodules, greater porosity, and, therefore, more aggregates in the compound. Thus, there is higher probability of aggregate contact and of smaller distances between aggregates, resulting in greater compound conductivity. Further, the higher DBP absorption of this grade signifying higher structure and bulkier aggregates also indicates a higher probability of aggregate contact or closer aggregate proximity, resulting again in greater conductivity. The lower surface area and DBP absorption values of N-294 and N-293, on the other hand, mean overall that there is less aggregate contact or proximity, fewer unhindered current paths, and thus a less conductive compound. Although both grades are essentially equal in "DBP," one can now predict that N-293 will be less conductive because of its lower surface area. This indeed is the case.

Carbon Black Grade and Loading

The selections of the proper carbon black grade and concentration are the two most important factors in determining the overall performance of a conductive plastics compound. Here the three carbon blacks, N-472, N-294, and N-293, were compounded into an ethylene–vinyl

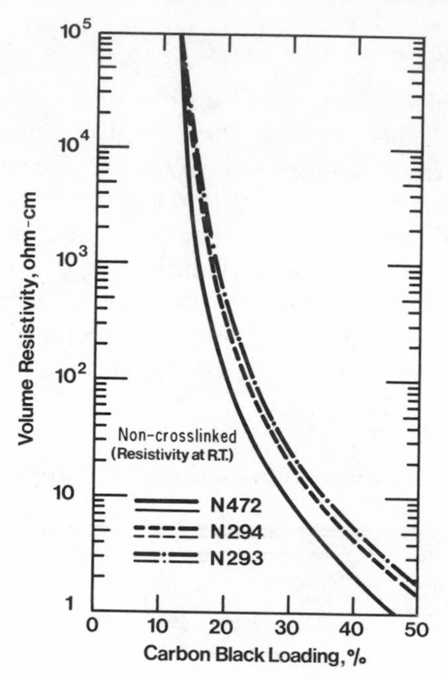

Figure 5. Effect of carbon black loading on resistivity (RT)

acetate (EVA) copolymer (Union Carbide DQD-1868, 19% vinyl acetate content) over a range of carbon black loadings. EVA was chosen for this work because it is widely used in the wire and cable industry for producing conductive compounds. Previous work has shown that in general, the relationships observed in this system are true for a wide range of thermoplastics (*7*).

The compounds were dispersed in a model B Banbury (1.2 liter capacity), using a fast dispersion cycle (~2½ min) and slow, medium, and high rotor speeds, and were converted to sheet form on a two-roll mill. Standard 80-mil tensile plaques were then molded using these millsheets, and 2″ × 6″-electrical test specimens were cut out of the tensile plaques. Each specimen was coated with a silver paint to produce a 0.5-inch wide silver electrode at each end. The specimens were placed in a sample holder, and the electrodes were attached to a Leeds and Northrup test set (No. 5305) consisting of a wheatstone bridge and galvanometer. The voltage impressed on the test specimens was approximately 4.5 V. The dc resistances across the length of the sample were measured and converted to volume resistivity in ohm-cm (*1*). The results of these tests are reported in Figure 5. The standard procedure for determining volume conductivity is to measure its reciprocal, volume resistivity. Therefore, in discussing the test results the term volume resistivity rather than volume conductivity will be used.

It is evident that carbon black grade N-472 imparts the lowest resistivity at any specific carbon black loading, followed by N-294 and

N-293 in that order. Figure 5 shows that compound resistivity decreases with increasing carbon black loading. This is consistent with the concept that the greater the number of aggregates in contact or separated by very small distances, the greater the conductivity of the resulting compound. At loadings between 15 and 20%, a marked decrease in resistivity occurs. With further increase in carbon black loading, resistivity decreases but at a much slower rate. With these carbon blacks, compounds containing 40% loadings or greater produce resistivities less than 10 ohm-cm, compared with resistivities on the order of 10^{14} to 10^{16} for pure unfilled polymers.

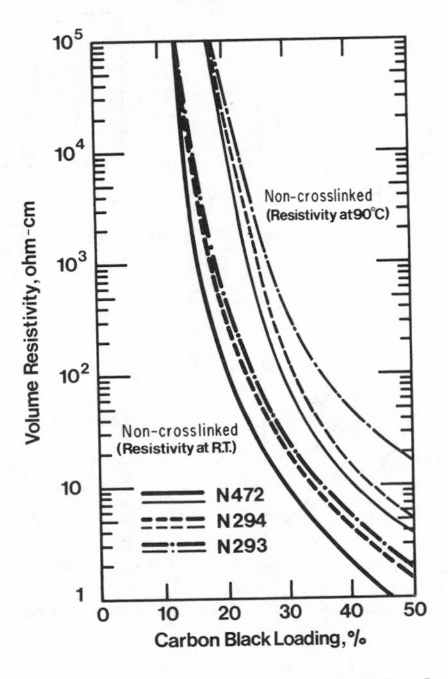

Figure 6. Effect of carbon black loading on resistivity (RT and 90°C)

Effect of Increasing Temperature on Resistivity

Specimens prepared as described above were also tested for volume resistivity at 90°C. As Figure 6 shows, at elevated temperatures carbon black-loaded compounds are higher in resistivity than at room temperature. A number of researchers have attributed this increase in resistivity to the expansion of the polymer phase, effectively reducing the carbon black concentration on a volume basis (*5*). Other workers relate this increase in resistivity to the crystalline nature of the polymers (*8*). They indicate that in crystalline systems, below the crystalline melting point the carbon black is not distributed uniformly through the polymer matrix but is concentrated in the non-crystalline or amorphous phase (8). They contend that upon heating the polymer above its crystalline melting point, the crystallites melt, at which time a portion of the carbon black in the amorphous phase migrates into the newly available area. This in effect reduces the carbon black concentration. Since the melting of the crystallites also sharply reduces the density of the compound, there is considerable difficulty separating the effects of lower density from that of crystallite melting.

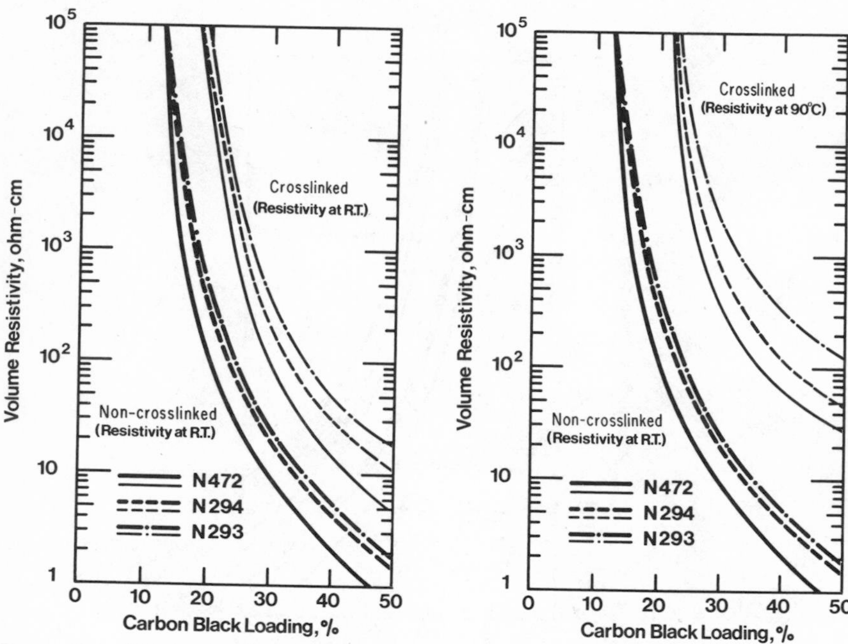

Figure 7. Effect of carbon black loading on resistivity (crosslinked and non-crosslinked)

Figure 8. Effect of carbon black loading on resistivity (RT and 90°C, crosslinked and non-crosslinked)

Effect of Crosslinking on Resistivity

Carbon black-loaded compounds prepared as described above were then compounded with a peroxide curing agent (Di-Cup T (α-dicumyl peroxide), Hercules, Inc.) at 2½% loading based on the polymer content, using a two-roll mill. The compounds were molded at 160°C so that they were crosslinked at the same time. Resistivity plaques were prepared as described above and tested at room temperature. Figure 7 shows the results of these tests; as can be seen, crosslinking increases the volume resistivity of the compounds. Since crosslinking reduces crystallinity and density, the resistivity increase may also be attributed to a reduction in the number of crystallites in the polymer as well as to a decrease in density.

The volume resistivity of these crosslinked compounds was then determined at 90°C (Figure 8). As might be expected, the effect of crosslinking coupled with increasing temperature increased resistivity still further. Apparently, the effects of further reducing crystallinity with increasing temperature as well as the decreasing of the density of the compound resulted in a still further increase in resistivity.

Effect of Carbon Black Loadings on Viscosity

Data presented thus far show that resistivity decreases with increased carbon black loading. Therefore, in those applications where very low resistivity is desired, compounds containing high concentrations of carbon black are needed. However, as more carbon black is added to a polymer system, the stiffness of the hot polymer mix is a controlling factor since extremely high viscosity formulations cannot be processed readily in either mixing or processing equipment, such as extruders. Figure 9 shows the effect of increasing carbon black concentration on the relative stiffness of ethylene–vinyl acetate compounds as indicated by meter-grams

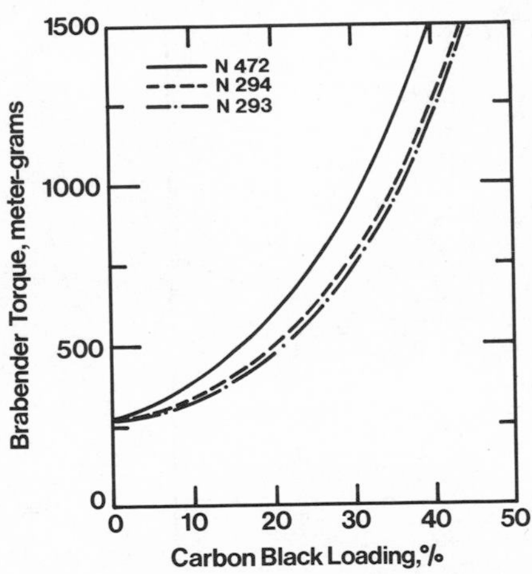

Figure 9. Effect of carbon black loading on viscosity

of torque. These determinations were made using a Brabender Plasticorder with a Roller 6, measuring head heated to 110°C. Previously prepared diced compounds were used in this study, and the torque readings were taken after 10 min of Plasticorder mixing to assure complete melting and uniformity.

As Figure 9 shows, all three carbon blacks significantly increase the viscosity or stiffness of the compounds as indicated by increasing torque over the range of 0 to 40% concentration. N-472, because of its higher surface area and DBP absorption, produces the highest viscosity, followed by the lower surface area and DBP grades, N-294 and N-293 in that order.

Physical Properties of Carbon Black-Loaded Polymers

Methods used to produce conductive polymers by the addition of conductive carbon blacks can result in plastics products with resistivity of less than 10 ohm-cm. However, conductive plastics products would be completely unsatisfactory unless they were able to meet the physical requirements of their end-use application. To show the effect on the physical properties of the polymer, the ultimate tensile and yield strength, elongation at break, and brittleness temperature of ethylene–vinyl acetate compounds were determined over a range of carbon black loadings using N-472. Ethylene–vinyl acetate was chosen for this work for the same reasons stated previously. N-472 was chosen because it is the most conductive of the carbon blacks. Again, both non-crosslinked and crosslinked compounds were tested. The tensile properties testing was per-

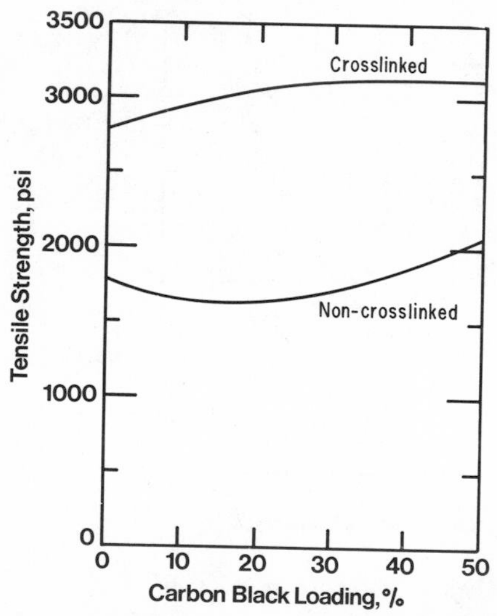

Figure 10. Effect of carbon black loading on tensile strength

formed using a Scott tensile tester according to ASTM procedure D-638 (rate of pull-20 inch/min). Brittleness temperature data were obtained using a Tinius-Olsen tester according to ASTM method D-746.

Tensile Strength. The effect of carbon black loading on tensile strength in ethylene–vinyl acetate polymer compounded with N-472 carbon black is shown in Figure 10. Both crosslinked and non-crosslinked compounds are reported. In non-crosslinked compounds low loadings of carbon black reduce tensile strength to a minimum point, and then further additions of carbon black produce a slight increase in tensile strength. Increasing loadings of carbon black in crosslinked compounds slowly increase tensile strength. In these compounds, the overall tensile strength of the compounds is significantly higher than that of non-crosslinked compounds. In general, however, carbon black in these systems does not increase or decrease tensile strength to a great extent over the range tested.

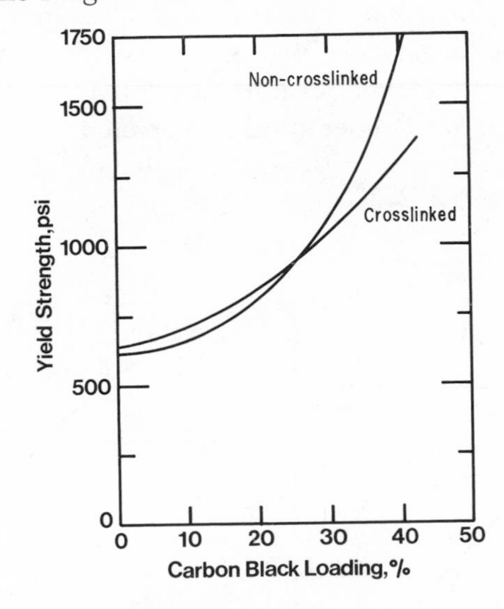

Figure 11. Effect of carbon black loading on yield strength

Yield Strength. Figure 11 shows the effect of carbon black loading on yield strength in an ethylene–vinyl acetate polymer compounded with N-472 carbon black. Both crosslinked and non-crosslinked compounds are reported. In both types carbon black addition increases yield strength significantly. At 30% loading and less, the effect of crosslinking on these compounds is very slight. However, at loadings higher than 30% the non-crosslinked compounds are higher in yield strength compared with the crosslinked compounds.

Elongation. Figure 12 shows the effect of carbon black loading on the elongation of an ethylene–vinyl acetate polymer compounded with N-472 carbon black. Again, both crosslinked and non-crosslinked compounds are reported. As shown, carbon black reduces elongation substantially in both. In this case, at carbon black loadings less than 30%, non-crosslinked compounds show greater elongations than crosslinked compounds. However, at loadings above 30% carbon black, the crosslinked compounds are higher in elongation compared with the non-crosslinked compound.

Elongation is one physical property that showed significant differences from one grade of carbon black to another. These differences, however, occurred only in the non-crosslinked compounds. As shown in Figure 13, N-294 and N-293 reduced elongation to the same degree while N-472 reduced elongation to a greater degree.

Low Temperature Properties. The low temperature properties of polymer systems are usually impaired by the addition of fillers such as carbon black. This can often be minimized by crosslinking the polymer system. Figure 14 shows the effect of carbon black loading on the brittleness temperature of both non-crosslinked and crosslinked ethylene–vinyl acetate compounds. Figure 14 indicates that carbon black loadings of 40% or more reduce the brittleness temperature of the polymers so that they could not be flexed at the low temperatures encountered in unusually cold outdoor exposures. Crosslinking of these formulations improves the brittleness temperature by about 25°C.

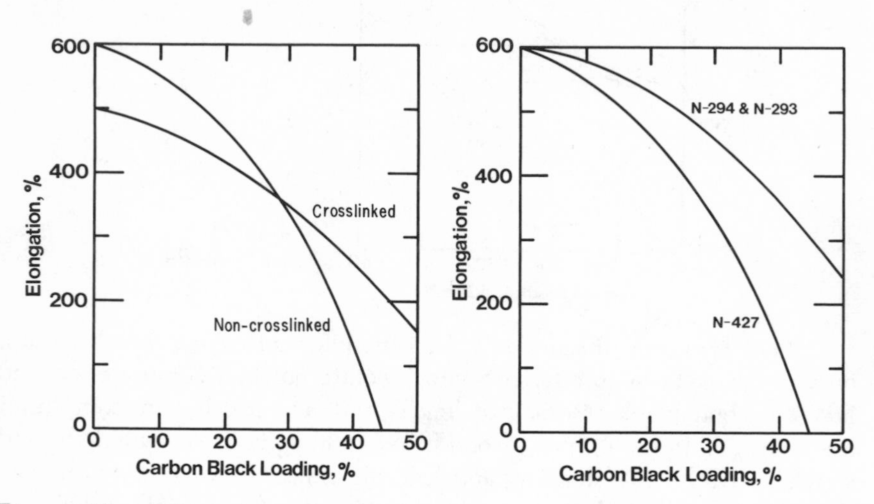

Figure 12. Effect of carbon black loading on elongation (crosslinked and non-crosslinked)

Figure 13. Effect of carbon black loading on elongation (N-294, N-293, N-427)

Conclusion

Resistivities of 10 ohm-cm or less can be achieved with thermoplastics at carbon black loadings of 40% or more. High loadings of carbon black increase viscosity and yield strength, reduce elongation, and reduce the serviceability of compounds at low temperature. The crosslinking of these compounds increases tensile strength and improves low temperature properties.

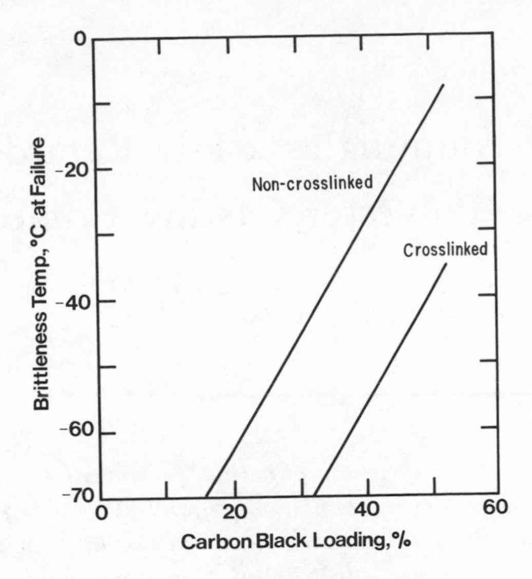

Figure 14. Effect of carbon black loading on brittleness temperature

Overall we hope that this work will assist compounders in selecting the carbon black grade and loading which will produce plastics compound combining the conductive and physical properties required for their application.

Literature Cited

1. Voet, A., *12th Annual Wire Cable Symp.*, Asbury Park, N. J., 1963.
2. Mildner, R. C., *IEEE Trans. Power App. System* (1970) **89**(2), 313.
3. Forster, E. O., *IEEE Trans. Power App. System* (1971) **90**(3), 913.
4. Polley, M. H., Boonstra, B. B. S. T., *Rubber Chem. Technol.* (1957) **30,** 170.
5. Van Beck, L. K. H., Van Pul, B. I. C. F., *J. Appl. Polymer Sci.* (1962) **6,** 651.
6. Riven, D., *Rubber Chem. Technol.* (1971) **40,** 307.
7. "Carbon Blacks for Conductive Plastics," *Tech. Bull.* **S-8,** Cabot Corp., Special Blacks Division, Billerica, Mass., 1970.
8. Mildner, R. C., Woodland, P. C., *13th Annual Wire Cable Symp.*, Asbury Park, N. J., 1964.
9. Norman, R. H., "Conductive Rubbers and Plastics," American Elsevier, New York, 1970.

RECEIVED October 11, 1973.

18

Hydrated Alumina as a Fire-Retardant Filler in Styrene–Polyester Casting Compounds

C. V. LUNDBERG

Bell Laboratories, Murray Hill, N. J. 07974

Hydrated alumina is an effective fire-retardant filler for styrene–polyester casting compounds. When equivalent volume loadings of hydrated alumina are used in place of silica and feldspar, self-extinguishing compounds are obtained when tested in accordance with ASTM D 635 in contrast to the combustible silica and feldspar compounds. The hydrated alumina compounds have oxygen index values ranging from 25 to 31, depending on the amount of filler present, compared with 20 for the silica and feldspar compounds and 21 to 22 for the latter compounds modified with 1.5 phr of antimony trioxide and 2.5 phr of chlorinated paraffin.

The work described here was started about three years ago when antimony trioxide, which is used as a fire retardant in styrene–polyester casting compounds, was in short supply and its price increased from 50¢/lb to more than $2.00/lb. Even at this price, suppliers were allocating it. One reason for the short supply, and possibly the main one, was that designers of plastic parts had become anxious to build flame retardance into their products, and the demand for antimony trioxide rose rapidly and exceeded the supply. As the price increased, other, probably less desirable flame retardants were adopted. The supply at the mines increased, and the price dropped so that today it is about 80¢/lb.

With the Nixon administration's policy of friendliness and trade with mainland China, a shortage of antimony trioxide may not occur again. China is a source which has not been tapped recently for political reasons but which should open up if the need arises. However, antimony trioxide, Sb_2O_3, often called antimony oxide, is still an expensive material volumewise since it has a specific gravity of 5.7. Thus, at 80¢/lb it has a pound-volume cost of $4.56. This compares with pound-volume costs of 36¢ for chlorinated paraffin, 22¢ for styrene–polyester resin, and 4¢ or less for mineral fillers such as feldspar and silica. Pound-volume cost is the product of the cost per pound and the specific gravity. It relates the cost to the volume; the denser the material, the less volume one pound of it will occupy.

The oxygen index test has become widely used in recent years to rate the flammability of plastics. Oxygen index is defined in ASTM D 2863 as "the minimum concentration of oxygen, expressed as percent by volume, in a mixture of oxygen and nitrogen which will just support combustion of a material under conditions of this method." This test is often referred to as the candle test since the specimen is ignited at its upper end similar to a candle.

The oxygen index test gives a numerical value which can be duplicated fairly closely from time to time in a given laboratory and between laboratories. An oxygen index of 28 has been adopted as the desirable minimum standard in our laboratories. However, no single test is satisfactory for establishing the flame retardance of materials under the variable conditions of an actual fire.

The use of 1.5 parts per hundred of resin (phr) of antimony trioxide together with 2.5 phr of chlorinated paraffin as a fire-retardant system for silica-filled styrene–polyester casting compounds was adopted at Bell Laboratories in 1963. Tests at that time were run on cast blocks measuring 3¼ inch × 1½ inch × 1 inch, which were similar in size and shape to 3A1A-3 protector blocks (apparatus which protects telephone equipment from electrical overloads). This protector block, made with the standard styrene–polyester casting compound in use at that time, failed to meet the Underwriters' Laboratories flammability requirement. The addition of 1.5 phr of antimony trioxide and 2.5 phr of chlorinated paraffin resulted in a protector block flame persistence of 18 seconds, well within the 60 seconds required by Underwriters. Their test is performed by applying to the bottom of a vertical specimen, at an angle of 20° to the horizontal, a flame 5 inches high with a 1½-inch blue cone, for five periods of 15 seconds duration with 15-second intervals. The antimony trioxide–chlorinated paraffin fire-retardant system was later adopted for use in virtually all Bell System styrene–polyester casting compounds.

When the oxygen index test became available, the styrene–polyester compounds with a flame persistence of 60 seconds or less in the Underwriters' test, had oxygen indices of 21 to 22, slightly above the percent oxygen in air (20.95). These compounds contained 1.5 phr of antimony trioxide and 2.5 phr of chlorinated paraffin. Thus, compound modifications seemed necessary to meet the required oxygen index minimum of 28.

Hydrated alumina has been mentioned in the literature at various times as a fire-retardant filler for rubbers and plastics. Thompson, Hagman, and Mueller (*1*) in 1958 and recently McCormack (*2*) reported on its effect in neoprene compounds. Martin and Price (*3*) and Stevens (*4*) reported on its effect in epoxy compounds. On the basis of favorable literature reports we decided to investigate the use of hydrated alumina as a fire-retardant filler in styrene–polyester casting compounds.

Compound and Specimen Preparation

Compounds were prepared in formula-weight or half-formula-weight batches based on a formula using 100 grams of resin. Mixing was done by hand in 8-ounce polyethylene-lined paper cups. The compounds minus the initiator and accelerator were evacuated at 3 mm of mercury, and Brookfield viscosities were run where appropriate. Initiator and accelerator were added, the material was again degassed at 3 mm of mercury and then cast in a Teflon-faced mold to a thickness of 1/8 inch. Thicker specimens (~3/8 inch) were cast in aluminum dishes with stainless steel electrodes for insulation resistance tests. Specimens were gelled for 2 hours at 125°F and cured for 2 hours at 250°F. The cured 1/8-inch thick sheets were cut to size for the various tests.

Table I. Filler Properties

	Feldspar 200 Mesh	*Silica 200 Mesh*	*Hydrated[a] Alumina*
Chemical Analysis, %			
loss on ignition	0.2	—	—
silica, SiO_2	71.5	99.8	0.022
alumina, Al_2O_3	16.3	—	64.9
iron oxide, Fe_2O_3	0.08 max.	0.03 max.	0.034
calcium oxide, CaO	0.4	—	—
magnesium oxide, MgO	trace	—	—
sodium oxide, Na_2O	4.0	—	0.22
potassium oxide, K_2O	7.5	—	—
Screen Test, %			
on 100 mesh	0.1	—	—
on 200	0.2	4	1–3
on 325	2.8	9	15–20
thru 325	96.7	80	80–85

[a] C-30BF alumina trihydrate, Alcoa.

Discussion

The first styrene–polyester compound made with hydrated alumina had a measured oxygen index of 26.4 compared with a silica–antimony trioxide–chlorinated paraffin control of 22.8, but it was more viscous than the control. The particular hydrated alumina used has a screen analysis showing 99% through a 325-mesh screen. A second grade of

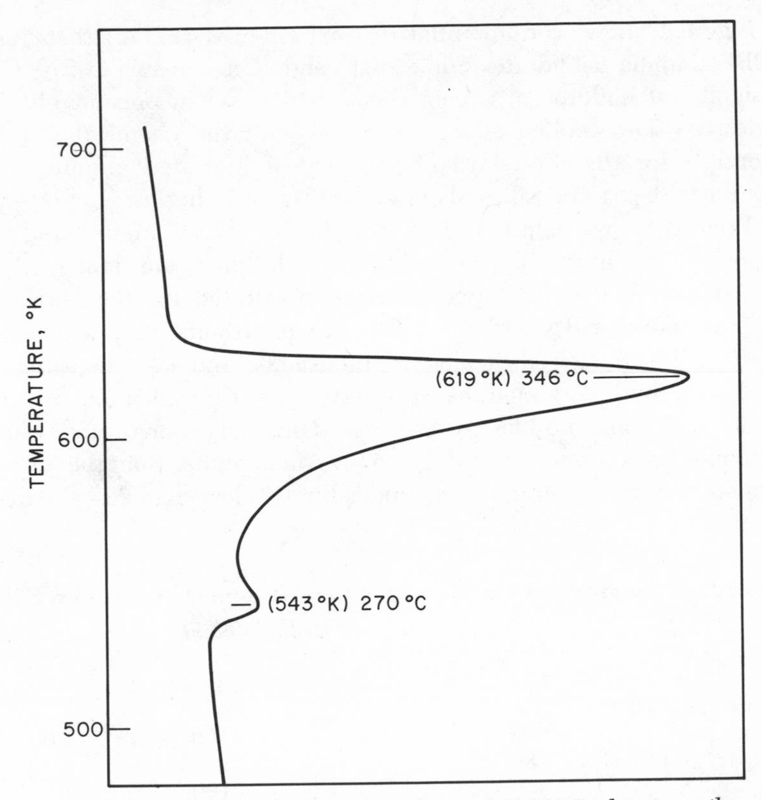

Figure 1. Differential thermal analysis of C-30BF alumina trihydrate (Al_2O_3–3 H_2O). Scan speed, 40°C/min; range, 64 (in nitrogen).

hydrated alumina with a screen analysis of 3–15% through a 325-mesh screen was then tried. The resulting compounds had oxygen indices of 26.0 and 25.4 for compounds with and without 2.5 phr of chlorinated paraffin, respectively. Even though this hydrated alumina is rougher in particle size than the first, it still resulted in compounds with viscosities greater than that of the silica control for equal filler volume loadings.

A third hydrated alumina resulted in favorable viscosities even though its sieve analysis showed that it had finer particle size than the second compound and its oil absorption (ASTM D-281) was close to that of the other two. All three hydrated aluminas are made by the Bayer process. The third contains a higher iron content and thus is less white than the others. Table I shows some of the properties of the third hydrated alumina (C-30BF alumina trihydrate, ALCOA), the one selected for the work described below, along with similar properties for the 200-mesh feldspar and the 200-mesh silica used for comparative purposes.

Figure 1 shows a differential thermal analysis (DTA) scan for the C-30BF alumina trihydrate. Some slight endotherm occurs at 270°C, but the significant endotherm is centered at 346°C when presumably water is released. The cooling effect associated with the chemical change is responsible for the fire-retardant property of hydrated alumina. DTA scans for feldspar and silica show no similar endotherms.

Table II shows data collected early in this investigation using (1) a compound containing 80 phr of feldspar plus fire retardants; (2) the same compound with hydrated alumina substituted for the feldspar on an equal-volume basis and no added fire retardants; (3) a compound containing 100 phr of silica plus fire retardants; and (4) the same compound with hydrated alumina substituted for the silica on an equal-volume basis and no added fire retardants. The hydrated alumina compounds have oxygen indices 3.7 to 5.3 points higher than the antimony trioxide–chlorinated paraffin compounds but still less than 28—the desired minimum.

Table II. Compounds Containing Equal Volume Loadings of Fillers

	Parts by Weight			
	T	*V*	*U*	*W*
Styrene–polyester resin[a]	100	100	100	100
Feldspar, 200 mesh	80[c]	—	—	—
Silica	—	—	100[d]	—
Hydrated alumina	—	73[c]	—	91[d]
Fiberglass, 1/32 inch	9.80	9.80	2.50	2.50
Titanium dioxide	0.46	0.46	—	—
Silicone antifoam	0.05	0.05	0.05	0.05
Cumene hydroperoxide	1.10	1.10	1.10	1.10
Cobalt octoate	0.25	0.25	0.25	0.25
Sb_2O_3	1.50	—	1.50	—
Chlor. par. (solid)[b]	2.50	—	—	—
Chlor. par. (liquid)[b]	—	—	2.50	—
Oxygen index	21.4	25.1	21.1	26.4

[a] Propylene–maleate–phthalate resin, 34.5% styrene.
Contains 70% chlorine.
[c] These quantities represent approximately equal volume loadings.
[d] These quantities represent equal volume loadings.

About the time that the above data were obtained, the styrene content of the styrene–polyester resin was increased from 34.5% to 38%. This resin change permitted an increase of the feldspar filler from 80 to 125 phr. It thus became desirable to obtain the oxygen index of the new

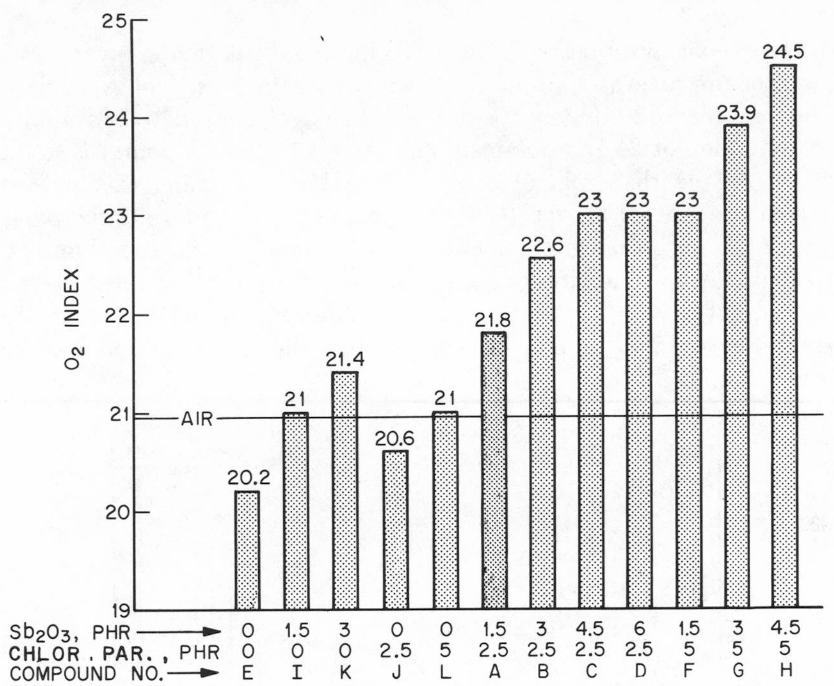

Figure 2. Effect of antimony trioxide and chlorinated paraffin in base compound E on oxygen index

compound and to determine whether increased quantities of antimony trioxide and/or chlorinated paraffin would result in an oxygen index of 28. Data for these compounds are presented in Figure 2; the formulation of the base compound is shown in Table III. The synergistic effect known to exist between antimony trioxide and chlorinated materials is obvious from the bar graphs. The chlorinated paraffin used in these compounds as well as in other compounds discussed contains 70% chlorine.

Table III. Base Compound E[a]

	Parts by Weight
Styrene–polyester resin[b]	100.0
Feldspar	125.0
Fiberglass, 1/32 inch	5.0
Titanium dioxide	0.5
Cumene hydroperoxide	1.0
Cobalt octoate	0.5

[a] This compound was modified by addition of antimony trioxide and/or chlorinated paraffin to produce compounds A,B,C,D,F,G,H,I,J,K,L.
[b] Propylene–maleate–phthalate resin, 38% styrene.

At a concentration of 2.5 phr of chlorinated paraffin, oxygen index does not improve as the antimony trioxide is increased above 4.5 phr. Using 4.5 phr of antimony trioxide and 5 phr of chlorinated paraffin, an oxygen index of 24.5 is obtained. The cost of this compound is greater (21.3¢/lb-vol) than that of one using hydrated alumina (20.4¢/lb-vol; oxygen index of 29.4), yet it does not meet the minimum desired oxygen index of 28. Increasing the amounts of antimony trioxide and chlorinated paraffin may increase the oxygen index to an acceptable level of 28 or greater, but it will be at a cost penalty. Also, increasing the chlorinated paraffin above 2.5 phr may adversely affect the electrical and corrosion properties.

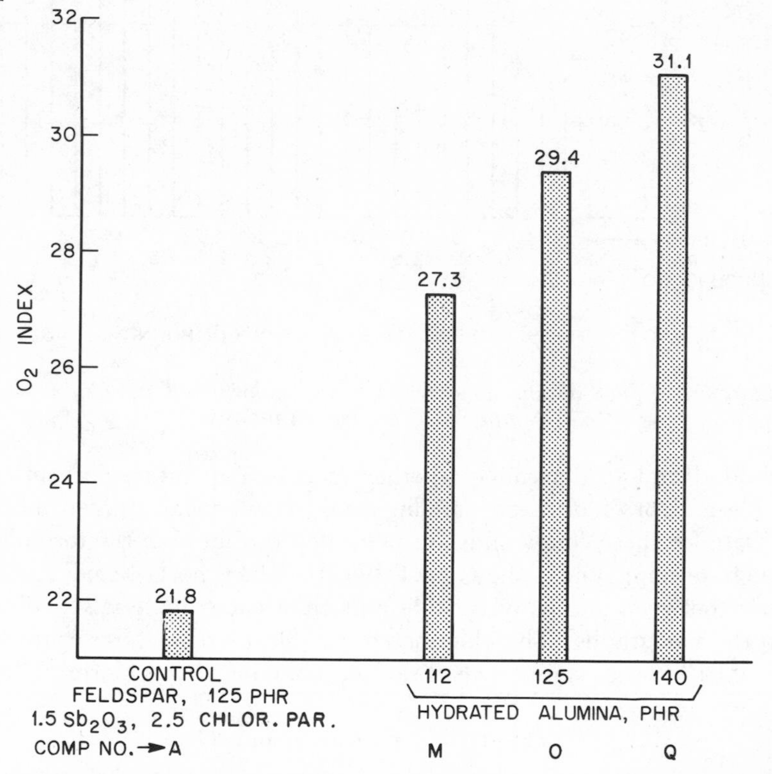

Figure 3. Oxygen index of styrene–polyester compounds filled with hydrated alumina

Figure 3 is a bar graph depicting the oxygen indices of styrene–polyester compounds filled with hydrated alumina compared with a feldspar–antimony trioxide–chlorinated paraffin control. The formulations of these hydrated alumina-filled compounds are shown in Table IV. The 112-phr hydrated alumina is equivalent in volume to 125 phr of feldspar. [The volume of each equals 46.3 (weight in phr/sp. gr. =

Table IV. Hydrated Alumina-Filled Compounds

	Parts by Weight				
	M	*O*	*Q*	*Y*	*Z*
Styrene–polyester resin[a]	100.0	100.0	100.0	100.0	100.0
Hydrated alumina	112.0	125.0	140.0	125.0	140.0
Fiberglass, 1/32 inch	5.0	5.0	5.0	2.5	2.5
Titanium dioxide	0.5	0.5	0.5	—	—
Cumene hydroperoxide	1.0	1.0	1.0	1.0	1.0
Cobalt octoate	0.5	0.5	0.5	0.5	0.5
Oxygen index	27.3	29.4	31.1	>28.[b]	>28.[b]

[a] Propylene–maleate–phthalate resin, 38% styrene.
[b] Estimated.

volume).] The oxygen indices of the compounds containing these amounts of fillers are 27.3 for the hydrated alumina compound and 21.8 for the feldspar–antimony trioxide–chlorinated paraffin compound. The viscosity of the 112-phr hydrated alumina compound is considerably less than for the feldspar compound so that more hydrated alumina can be tolerated. Increasing the hydrated alumina to 125 phr (volume of 51.7) increases the oxygen index to 29.4, and increasing to 140 phr (volume of 57.9) increases the oxygen index to 31.1. The viscosity of this latter compound is comparable with that of the 125-phr feldspar compound.

Table V compares flammability as measured by the oxygen index method with flammability as measured by ASTM D 635, "Flammability of Self-Supporting Plastics." In this test the specimen (5 inches × ½ inch

Table V. Oxygen Index *vs.* ASTM D 635 Flammability Ratings

Compound	*Resin*[a]	*Filler*	*Loading*		*ASTM D 2863*	*ASTM D 635 Burning*
			phr	*vol*	O_2 *Index*	*Time, Sec*[b]
T[c]	1	feldspar	80	29.6	21.4	Burns > 180
A[c]	2	feldspar	125	46.3	21.8	Burns > 180
H[d]	2	feldspar	125	46.3	24.5	30
U[c]	1	silica	100	37.7	21.1	Burns > 180
V	1	hyd. alumina	73	30.2	25.1	115
W	1	hyd. alumina	91	37.6	26.4	45
M	2	hyd. alumina	112	46.3	27.3	20
O	2	hyd. alumina	125	51.7	29.4	0
Q	2	hyd. alumina	140	57.9	31.1	0

[a] Resins differ only in styrene content: resin 1 contains 34.5% styrene; resin 2 contains 38% styrene.
[b] Time for the burning specimen to extinguish after removal of the burner flame from the specimen, officially called "Average Time to Self-Extinguishment." Specimen thickness 1/8 inch.
[c] Contains 1.5 phr Sb_2O_3 + 2.5 phr chlorinated paraffin.
[d] Contains 4.5 phr Sb_2O_3 + 5 phr chlorinated paraffin.

× ⅛ inch) is supported in the horizontal position with its transverse axis (width) inclined at 45° to the horizontal. A 1-inch blue Bunsen burner flame is applied to the free end of the specimen for 30 seconds. The time for the burning specimen to extinguish after removal of the burner flame from the specimen is measured in seconds. In this test the feldspar and silica compounds containing 1.5 phr antimony trioxide and 2.5 phr chlorinated paraffin burn. Increasing the flame retardants in the feldspar compound to 4.5 antimony trioxide and 5 chlorinated paraffin (Compound H) produces a compound which extinguishes in 30 seconds (oxygen index of 24.5). Each increase in the loading of hydrated alumina results in a decrease in the burning time and a corresponding increase in the oxygen index.

Table VI shows the mechanical properties of the hydrated alumina compounds *vs.* the feldspar–antimony trioxide–chlorinated paraffin control. The viscosities previously referred to are shown numerically. The mechanical properties for the feldspar and the three hydrated alumina compounds are similar. The electrical properties of the hydrated alumina compounds are comparable with those of the feldspar compound.

An important characteristic of any filler used in a liquid system is its settling properties during storage of the mixed compound before final processing and cure. The hydrated alumina settles more tightly on prolonged standing than the feldspar and silica fillers. This may not be a serious problem when the styrene–polyester compounds are mixed where they are used, and thus storage time of mixed material is short.

Compound Costs

The costs of the compounds are shown in Table VII and are based on the ingredient prices shown in Table VIII. Costs are shown based on the F.O.B. prices at the mines or point of manufacture. The calculated compound costs are satisfactory for comparison but may not be exact.

Table VI. Mechanical Properties of Hydrated Alumina Compounds

Properties	*Compounds*			
	A (*Control*)	*M*	*O*	*Q*
Hydrated alumina, phr	—	112	125	140
Viscosity, Brookfield, cps, No. 3 spindle, 4 rpm	15,250	5,870	9,250	17,000
Oxygen index, %	21.8	27.3	29.4	31.1
Hardness, Barcol	65	60	62	63
Modulus of rupture, psi	9,600	10,200	9,790	9,060
Modulus of elasticity, psi × 10^{-6}	1.58	1.37	1.46	1.52
Deflection temperature, °C	95	96	98	99

Table VII shows that use of hydrated alumina at the 125-phr level (oxygen index of 29.4) to replace feldspar increases the compound pound-volume cost by approximately 5%. The corresponding compound cost increase when hydrated alumina is used to replace silica is 3%. As previously noted, increasing the antimony trioxide–chlorinated paraffin to 4.5 and 5 phr, respectively, in the feldspar compound, increases the pound-volume cost by approximately 10%; hence, this feldspar com-

Table VII. Compound Costs[a]

	Compound	*Sp. Gr.*	*Cost per lb, ¢*	*Cost per lb-vol, ¢*	*% Incr. over A*[b]
A	125 phr feldspar with F.R.[c]	1.66	11.7	19.4	—
H	125 phr feldspar with F.R.[d]	1.68	12.7	21.3	9.8
O	125 phr hydrated alumina	1.59	12.8	20.4	5.2
Q	140 phr hydrated alumina	1.63	12.3	20.1	3.6
					% Incr. over U[b]
U	100 phr silica with F.R.[c]	1.57	12.0	18.8	—
W	91 phr hydrated alumina	1.49	13.5	20.1	7
Y	125 phr hydrated alumina	1.59	12.2	19.4	3
Z	140 phr hydrated alumina	1.62	11.7	19.0	1

[a] Without cumene hydroperoxide and cobalt octoate.
[b] Cost per lb-vol % increase over A or U.
[c] 1.5 phr Sb_2O_3 + 2.5 phr chlorinated paraffin.
[d] 4.5 phr Sb_2O_3 + 5 phr chlorinated paraffin.

Table VIII. Prices of Ingredients Used in Calculating Compound Costs

Material	*Sp. Gr.*	*¢/lb*	*¢/lb-vol*
Antimony trioxide	5.7	79.5	452.3
Chlorinated paraffin, solid	1.66	22.5	37.4
Chlorinated paraffin, liquid	1.56	23.0	35.9
Fiberglass	2.54	66.0	167.7
Styrene–polyester resin	1.10	20.5	22.4
Titanium dioxide	3.88	26.73	104.0
Fillers, F.O.B., bagged, carload:			
hydrated alumina	2.42	4.45	10.77
feldspar, 200 mesh	2.70	1.375	3.71
silica, 200 mesh	2.65	0.775	2.05

pound is more costly than the hydrated alumina compound which has an oxygen index of 29.4, and yet the feldspar compound has an oxygen index of only 24.5. Thus, the more expensive filler—hydrated alumina—results in a compound with an oxygen index of 28 minimum and at a lower cost than for a feldspar–antimony trioxide–chlorinated paraffin compound.

Safety

Employee exposure to silica dust can result in silicosis, a disease of the lungs. Good ventilation and protective equipment are required for the safe handling of silica. Aside from a company's high standards for safety and health, OSHA regulations may become more restrictive in the future with regard to silica. Hydrated alumina is not known to cause lung illness similar to silicosis, and its use requires only the normal precautions which should be followed when handling mineral dusts.

Conclusions

The use of antimony trioxide combined with chlorinated paraffin in styrene–polyester casting compounds does not result in sufficient flame retardance to have an oxygen index of at least 28% and still be the most economical. This requirement can be met by using hydrated alumina as filler in place of mineral fillers such as feldspar and silica and omitting the antimony trioxide and chlorinated paraffin. The electrical and mechanical properties of compounds containing hydrated alumina are in line with the properties obtained with feldspar and silica compounds. Increasing the antimony trioxide and chlorinated paraffin above 1.5 and 2.5 phr, respectively, in feldspar and silica compounds increases the oxygen index, but the maximum amounts of these materials used in this study resulted in an oxygen index of 24.5—considerably below the desired 28—and the pound-volume cost of this compound is higher than that of hydrated alumina compounds having oxygen index values greater than 28.

Acknowledgment

The author thanks E. W. Anderson and R. J. Caroselli for running the electrical tests, and F. X. Ventrice for running some of the physical tests.

Literature Cited

1. Thompson, D. C., Hagman, J. F., Mueller, N. N., *Rubber Age* (Aug. 1958) **83,** 819-824.
2. McCormack, C. E., *Rubber Age* (June 1972) **104,** 27-36.
3. Martin, F. J., Price, K. R., *J. Appl. Poly. Sci.* (1968) **12,** 143.
4. Stevens, J. J. Jr., "Improved Cycloaliphatic Epoxy Systems for High Voltage Applications," *Proc. Electrical Insulation Conf. IEEE, 9th* (Sept. 10, 1969).

RECEIVED October 11, 1973.

19

Corrosion Engineering in Reinforced Plastics

OTTO H. FENNER

Monsanto Co., 800 N. Lindbergh Blvd., St. Louis, Mo. 63166

Reinforced plastics provide the materials engineer with a solution to the economic dilemma arising from the current prohibitive cost of exotic alloy construction. They are serving in highly corrosive environments as materials for many major pieces of process equipment, auxiliaries, and structural members. The excellent chemical resistance properties of FRP, epoxy, furan, phenolic, and polyester resins make each a premium performer in specific services. "One-side" panel testing, along with changes in residual flexural strength of "dunk" test strips, appears to be the best way to evaluate the performance of plastics. Failures in laminated plastics are readily solved with the electron scanning microscope through plastographic analysis techniques and infrared spectroscopy.

Successful corrosion engineering depends to a great deal on an in-depth knowledge of chemistry and chemical reactions. The chemist or process engineer engaged in materials selection for construction of equipment destined for use in highly aggressive environments must first be well founded in both inorganic and organic chemistry. This is particularly true when it becomes necessary to use an intuitive choice of a material of construction for lack of specific corrosion resistance data. The chemical resistance and probable performance of non-metallic materials under such conditions may be readily and often accurately predicted by considering the chemical structure of the compounds to be encountered in the environment. It is because of his extensive, specialized training in the fields of inorganic, organic, and physical chemistry that we find the chemist is often successful in industry as a corrosion engineer.

During the past 20 years there has been a tremendous growth in the use of reinforced plastics for fabricating process equipment. Their fine performance, within the limits of their temperature and physical properties, has been demonstrated repeatedly in application. Plastics no longer need to be regarded with hesitancy or trepidation when considered with respect to the costly exotic alloys and metals. They have experienced extremely satisfactory service in many highly corrosive environments.

Through the ingenuity of the polymer chemist the properties of plastics can be altered through formulation and the incorporation of specific fillers and reinforcements to obtain modifications approaching the strength of steel, the lightness of aluminum, and a chemical resistance better than some of the super alloys. This is a prime reason for their phenomenal acceptance in chemical processing equipment.

Plastic materials are highly important to the cost-conscious maintenance and design engineer. Generally they have relatively low initial costs, are lightweight and easily installed in hard-to-reach areas, and can often be repaired in place without the prohibitive costs of shutting down adjacent equipment. In most instances the expense of post-fabrication stress-relieving heat treatments, often associated with the construction of alloy equipment, is unnecessary.

Except for specifically formulated and compounded plastics or resin systems, these materials do not conduct electrical current. Thus, one of the chief concerns of the corrosion engineer—galvanic and stray current corrosion—is non-existent when such materials are used.

Color pigmentation during the fabrication of a reinforced plastic unit permits an extra aesthetic premium not available in metallic structures requiring post-construction painting. Quite frequently metal-ion contamination leads to serious color and taste problems during the manufacture of pharmaceutical or food grade compounds in metal or alloy equipment. This is a very mild type of attack referred to as "micro-mil" corrosion. Stainless steel alloys have been used in many instances where such contamination or quality control is the problem rather than severe, prohibitive corrosion. Often even stainless steel is inadequate, and reinforced plastics have been required as substitutes.

Because of the serious quality problems arising from this miniscule amount of corrosion, the materials or corrosion engineer often is called upon to assist the production supervisor in recommending an alternative material of construction. Although this minute attack will introduce metal ions in parts per million into the product, this is frequently significant enough to impart adverse color effects or undesirable tastes to not only the main compound but subsequent products. The use of reinforced plastics often has solved the problem.

In many instances no metal or alloy is either practical or economical to use for fabricating a custom-built piece of equipment which will be exposed to a highly corrosive environment. This is certainly true in many applications dealing with strong oxidizing acids, wide ranges of pH, and acidic salt solutions. In these applications, specific reinforced plastics are invaluable.

By using plastic structures with a high degree of translucency, safer and easier operations are encountered, as well as savings of thousands of dollars for liquid level-measuring auxilliaries. The ability to determine visually the height of a liquid in a tank at a glance lessens the hazard of spillovers into an operating area.

One property of plastic materials of construction which must be considered from a safety viewpoint is their ability to build up a static charge. Rapid bleed-off and sparking can be particularly hazardous around flammable liquids or dry dust environments. Suitable grounding procedures or neutralizing media must be used in these situations to guarantee safe working conditions. Since many plastics and reinforced resins will burn, fire-retardant steps are necessary whenever the question of safety arises from this source. Generally the solution is simply the modification of the resin formulation by the addition of fire-retardant compounds or by the internal modification of the resin molecule itself.

From an engineering standpoint, the most important design consideration is the structural soundness of the unit. This means that the lay-up and thickness of the laminate must be evaluated to make certain that not only will the structure withstand all of the hydraulic or static load pressure impressed upon it but also that any proposed vacuum or pressure service conditions will be considered. The unit should be fabricated to withstand any vibrational fatigue arising from agitation of fluids or heavy slurries within the tank. Any agitator drive assembly must be taken into consideration. This should be externally self-supported when the tank walls are not sufficiently strong and thick enough to support the added weight.

In the majority of cases, failures in glass fiber reinforced plastics are physical-mechanical in nature. In the construction of such equipment, the following cause the most frequent problems:

(1) Inadequate wetting of the glass fibers by the resin, resulting in dry spots and delaminations

(2) Entrapped air, resulting in subsurface air bubbles and blisters or craters

(3) Non-uniformity of laminal construction, such as fold-overs and irregularities of laminal thickness

(4) Prohibitive porosity and crevices at junctures of nozzles and tank wall

(5) Inadequate reinforcement and/or support of nozzles and tank walls

(6) Exposed glass fibers of the reinforcing mat to the corrosive environment because of missing surfacing wall

(7) Insufficient thickness in protective surface gel coat

(8) Application of gel coating over a paraffin-waxed surface, resulting in poor adhesion

(9) Excessively rich resin areas which produce brittle sections, shrinkage cracks, and high temperature exotherms

(10) Internal stress cracking

(11) Unfinished sections, unreinforced areas, and careless workmanship

(12) Prohibitively thin flanges, leading to warped faces and distortion

(13) Contamination from foreign materials in the working area, embedding within or upon surface of resin

In addition to the above, the curing procedure must be studied to determine whether it is adequate or not. An improperly cured unit can give disastrous results in service.

Reinforced Plastics

Epoxy. The reaction products of bisphenol A and epichlorohydrin, known as epoxy resins, have excellent adhesive qualities and exhibit good chemical resistance. They can be readily cured by catalysts and accelerators at ambient temperatures, which makes them extremely worthwhile in patching process equipment or reinforcing fragile materials. Generally reinforced with glass fibers, the epoxies show very good resistance to water, sour crudes, salt water, strong alkaline solutions, nonoxidizing acids, jet fuels, aliphatics, and some aromatic compounds. The generally recommended temperature for maximum operations is about 250°F. However, good results have been experienced with the glass fiber-reinforced epoxy as high as 350°F in acidic vapors of pH 3.

The epoxies have been used for glass fiber-reinforced storage tanks, receivers, feed tanks, columns, scrubbers, and ducts for handling highly corrosive liquids and vapors; protection of fragile materials such as glass piping, ceramics, stoneware, carbon, and graphite from breakage under tensile stresses through armouring with glass tape and epoxy resin adhesive overlay; protective coatings and linings for tanks, storage bins, hoppers, and chute-handling wet crystals and dry powders; monolithic floor toppings to promote acid-resistant floors and mortars for acid-proof brick floor.

Furan. The fully cured furan resins have excellent chemical resistance to strong acids, concentrated alkalies, chlorinated solvents, and moisture. They are not resistant to such oxidizing agents as nitric acid, chromic acid, concentrated sulfuric acid, chlorine, and hypochlorites. Their high degree of chemical resistance to chlorinated hydrocarbons such as chloroform, chlorobenzene, and dinitrochlorobenze has made the furans a premium material for protection of equipment where few other materials are unaffected. Although the recommended continuous operating temperature for the glass fiber-reinforced furans is 280°F, successful applications as high as 350°F have been reported. The furan resins have been used extensively for the manufacture of pipes, ducts, valves, pumps, blowers, jets, exhaust and vent stacks, scrubbing towers, receivers, distillation towers, tanks, mist eliminators, absorbers, heat exchangers, filters, presses, pressure filter tanks, tank linings, and acid-proof brick mortars.

Phenolics. The phenolics have been performing yeoman chores in protecting chemical equipment for nearly 50 years. These heat-cured resins are insoluble in most solvents. They have high acid resistance and are inert to most of the common chemicals. Their resistance to moisture is very good. In general, the unmodified phenolics exhibit poor resistance to alkaline media as well as to strong oxidizing agents. These materials have been used at temperatures as high as 350°F for prolonged periods. Their dimensional stability remains quite high over a wide temperature range. The phenolics are classified as non-flammable materials.

The reinforced phenolics may use carbon or graphite, asbestos, glass fiber, or natural and synthetic fibers. They are used to fabricate process piping, valves, pumps, blowers, and tanks. Their applications include asbestos-filled phenolic resin crystallizers, quenchers, boiling tubs, hold tubs, storage and feed bins, hoppers, stacks, and ducting; phenolic resin-impregnated graphite valves, pumps, jets, heat-exchangers, distillation (bubble-cap or packed) columns, and split-flow controllers; filters, presses, coated tray and shelf vacuum dryers, rotary dryers, and pressure and vacuum filter tanks; fume and acid vapors scrubbing towers and exhaust jets.

Polyesters. The chemical-resistant polyesters include the well known propoxylated bisphenol A—fumarate resin; one based upon hydrogenated bisphenol A and fumaric acid; the isophthalates; and polyesters made with chlorendic anhydride or tetrachlorophthalic anhydride. These resins have good chemical resistance to non-oxidizing acids, corrosive salts, aliphatic solvents, aromatic compounds, and chlorinated solvents. They exhibit fair resistance to weak bases and esters but are severely attacked by strong oxidizing acids and strong bases.

The polyesters are generally reinforced with glass cloth, mat, rovings, chopped fibers, or glass flakes in the fabrication of chemical processing equipment. Although the glass fiber-reinforced polyesters are usually employed at continuous operation temperatures below 212°F, they have been successfully applied in strongly corrosive environments as high as 250°F. Special modifications with reactive diluents or cross-linking monomers such as triallyl cyanurate have been reported with heat stability at temperatures as high as 500°F.

The glass fiber-reinforced polyesters have exhibited a versatility of use which ranges from parts of the lunar excursion module to mine sweepers in Vietnam waters. Their application in the chemical industry includes reactors, storage tanks, catch tanks, boiling tubs, ground tanks, receivers, and separators; reclaimed water sumps, funnels, laboratory sinks; ducts, hoods, and vent stacks; dip pipes, process piping, sewers; dry powder bins, feeders, chutes, carts; wet cake hoppers, screw-feed troughs, conveyor covers; chlorine absorbers, off-gas scrubbers, mist eliminators, and columns; doors, windows, canopies, paneling, structural members; equipment safety shields, motor covers; fans, blowers, eductors, jets, exhausters; filter press frames, slide blast gates, check valves; waste disposal troughs, settling basins, cooling towers, guttering, downspouts; chopped fiber spray and glass flake linings for in-place rehabilitation of process equipment.

Testing

The testing of glass fiber-reinforced plastics either in the laboratory or under actual plant operations by the simple immersion method is often sufficient to determine acceptability by the environment. This is particularly true when the plastic is readily attacked by the test solution. However, when there is little visible change in the test sample, determination of its compatibility with the particular environment becomes somewhat more difficult. It is possible, for an example, that undesirable trace contaminants present in the process stream may be absorbed over a period of time or permeate the resin with no immediate visible evidence of harmful effects. A gradual degradation of the physical and mechanical properties of the plastic may be developing a weakened structure. Under those conditions of no apparent visible change, it is generally best to make certain that no insidious or adverse conditioning of the plastic is occurring which could promote future difficulties. Thus, performance data based upon changes in physical and mechanical properties are very meaningful (*1*).

Initially, testing techniques similar to those used in evaluating the corrosion resistance of metals and alloys may be used for the glass fiber-reinforced resins. The specific test coupons are exposed to the environment either in the laboratory under as nearly as possible simulated plant conditions or under actual field operations in the process stream. The coupons are usually positioned to provide both liquid- and vapor-phase environmental exposures.

When exposure is finished, the test coupons are removed and examined for changes in weight, dimension, appearance, hardness, delamination, physical and/or mechanical properties, and actual dissolution of the plastic.

Incompatibility of a reinforced plastic with the environment is visually apparent by such occurrences as those described below.

Abnormal Surface Changes. These may be identified by charring, chalking, blistering, cracking, and crazing, all of which can result in a weakened structure and prohibit the extended use in the particular service.

Color Changes. While indicating some interaction of the plastic and the environment, such changes do not necessarily prohibit the use of the plastic. Many times this color change is limited either to the immediate surface or at worst to a few mils penetration. Permeation sometimes causes the development of a color change throughout the body of the plastic. However, often the discolored plastic will not have experienced any significant degradation in its physical or mechanical properties and is entirely acceptable in the proposed service.

Dimensional Changes. If swelling or shrinkage is minor, it is no reason for rejection provided such dimensional changes do not prohibit the plastic from performing its function. Minor swelling in the plastic lining of a tank is not sufficient cause for alarm or replacement as long as its adhesion to the tank wall is unaffected. Swelling to the same extent within the volute of a pump, body of a valve, or throat of a flow metering device might be quite unsatisfactory. Prohibitive swelling is often only in the eyes of the evaluator and quite difficult to define in terms of percentage growth as to rejection or acceptance. On the other hand excessive shrinkage is more easily defined by shrinking, cracking, distortion, or loss in adhesive qualities.

Hardness Changes. These are not uncommon. Softening or hardening of a plastic or resin system exposed to a particular environment may or may not be evidence of prohibitive attack. Softening, in many instances, is solely the result of the permeation of a solvent into the resin with some attendent plasticizing. Upon removal from exposure to the environment, the solvent may evaporate and the plastic may resume its original hardness. Tackiness accompanying a softening of the resin system generally denotes its unacceptability. However, this may be only a surface phenomenon and a reasonable, practical life could be experienced by the material in the environment. Brittleness and losses in tensile or flexural strength often are associated with surface hardening of the plastic. Certain media, such as strong acids, may increase the surface hardness of catalytically cured resins without detrimental effects.

Laminal Wall Attack. This is suffered by fiber-reinforced resin structures in environments in which the resin is unacceptable (*1*). However, should the resin system not have been properly cured prior to exposure, failures might occur in an otherwise acceptable environment. In other instances, delamination occurs because there is inadequate wetting of the glass, resulting in poor adhesion. The failure to remove paraffin wax from the surface of a glass fiber-reinforced polyester laminate prior to the addition of any overlay can cause delamination too. Exposure of glass fibers to environments such as strong bases or hydrofluoric acid which are incompatible with glass can cause rapid deterioration of the glass reinforcement and delamination.

Weight Changes. These are used by many laboratories in reporting the performance of plastics. These weight changes result from the absorption of liquids, production of corrosion products, leaching of resin, erosion, corrosion, or other mechanical effects, deterioration of reinforcing fiber through chemical attack, or a combination of these. Often a single weight measurement at the end of the exposure period is sufficient for evaluating the plastic for the specified environment. However, false or questionable conclusions or inferences may result from relying solely on such measurements. To avoid this, a series of physical/mechanical property tests and weight change evaluations should be made on each plastic at definite time interval during the exposure period. These measurements are plotted so that a projected performance life may be predicted from the curves. A fairly satisfactory technique plots the accumulated percentage weight change from initial weight along with the differential percentage weight change between successive weighings. This combination will indicate whether the rate of change is increasing, decreasing, or remaining constant for any increment of time during exposure to the aggressive environment.

Residual Physical Properties Method. This is probably the most important consideration in the evaluation of glass fiber-reinforced plastic laminates which have been subjected to severely corrosive environments. Whether they have retained sufficient amounts of their original compressive, flexural, and tensile strengths to provide a structurally sound unit over a reasonable service life is of paramount importance.

Review of many test exposures of glass fiber-reinforced resin samples discloses that both initial loss or gain in weight, or even a decrease in flexural strength or modulus is only part of the performance picture. Rate of change or even stability of these with additional exposure is much more a criterion for acceptance. Many investigations disclosed rapid loss in physical properties the first few hours or days of exposure, with a subsequent flattening out of the curves which indicate no appreciable further lowering of the strength of the laminate.

These studies lead to the conclusion that changes in flexural strength and flexural modulus are the best criteria for evaluating the chemical resistance of glass fiber-reinforced polyesters both in plant and in laboratory exposures. Briefly, the test procedure is based on the evaluation of changes in flexural strength or fluxural modulus of the exposed samples with respect to time. These determinations are used as the standard reference between performance of glass fiber-reinforced polyesters since changes in these properties appear to be the most indicative check on the physical behavior of these materials.

Deflectron Test Cell. It has become quite apparent to many research chemists and materials engineers that the standard "dunk" test introduces difficulties in interpretation when attempting to use the data for good engineering design. In considering the effects of cold walls upon the performance of materials, almost simultaneously, many companies and

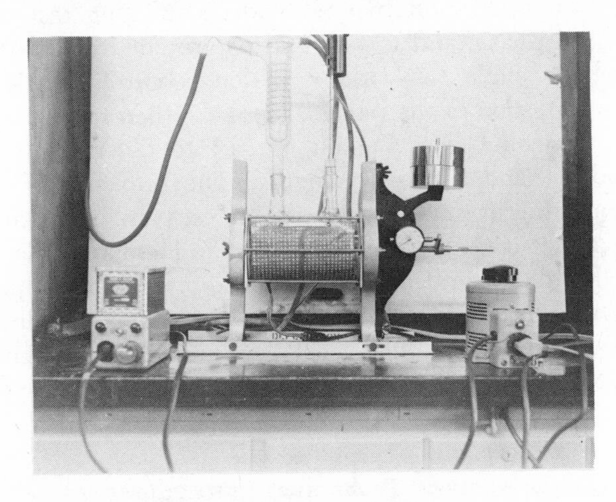

Figure 1. Open end test cell

personnel interested in this field have arrived at the decision that "one-side" testing of the glass fiber-reinforced plastic panel under conditions simulating the inner wall of a tank or duct is more desirable.

Monsanto Corporate Engineering uses the Deflectron open end test cell (Deflectron Co., St. Louis, Mo.) to obtain realistic exposure data in the evaluation of glass fiber-reinforced polyesters and other sheet plastics. The test apparatus (Figure 1) consists of a 4″-diameter × 10″-long glass cylinder with several ground-glass joint openings placed in line perpendicular to the axis of the cylinder. The ends of the glass cylinder are closed off with the resin glass laminate to be tested and sealed with suitable gaskets, allowing one-side exposure only to the particular environment. The cell is half-filled with the test solution which provides an equal liquid phase–vapor phase exposure area upon each end test panel. The unit can be fitted with the desired auxiliary equipment such as reflux condenser, thermometer, thermostatic control, and agitator. The environmental cell is heated with an electric heating mantle to maintain the required temperature.

In addition to the 6″ × 6″ × ⅛″-thick side test panel, generally ½″ × 3½″ × ⅛″-thick sealed edge "dunk" test strips are immersed in the liquid. One of these is removed after 1, 3, 7, 14, 30, 60, 90, and 120 days to observe any visible attack over the period of immersion. Changes in Barcol hardness are determined, and then residual flexural modulus and flexural strength determinations are made on each test strip upon its removal, according to ASTM D-790, Standard Method for Test for Flexural Properties of Plastics.

Upon completion of the testing period, usually a 120-day exposure, the unit is dismantled, and two ½″-wide × 6″-long test strips are cut from the liquid phase, and two test strips are cut from the vapor phase of the panel. A single ½″-wide × 6″-long control sample is cut from the unexposed portion of the panel. These are then broken on an Instron instrument (Instron Corp., Canton, Mass.) to determine their residual flexural modulus and flexural strengths. The broken test strips are reassembled into their former position in the test panel for future reference.

The fact that control panels of a specific laminated plastic structure may be used with a number of Deflectron test cells to compare chemical resistance with those of other resin laminate systems is a decided advantage. The variation in chemical resistance through differences introduced by various reinforcing fibers such as asbestos, ceramic, nylon, acrylics, polyesters, or other synthetics is often of importance and has been evaluated by this technique.

Plastographic Analysis Techniques. Procedures for obtaining additional performance data after plastic materials have been subjected to aggressive environments, termed plastographic analysis techniques (*2*), are often used. The technique has been extremely helpful in performance and failure analyses studies which have led to the development of better engineering materials of construction and design considerations for plastic and reinforced materials of construction.

In this procedure plastic samples and glass fiber-reinforced resin laminates are prepared and subjected to "known conditions" of exposure. We have used it most frequently in the permanence studies with fiber-reinforced resin materials. In these studies typical "known condition" samples for comparison against SPI standard control laminates were prepared by the following methods:

(a) Physical fractures of the SPI standard control laminate through fold-over, torsional breakage, flexural fatigue cracking, and hydrostatic rupture

(b) Undercured laminates arising from insufficient additions of catalysts or accelerators, excessively low cure temperatures, air inhibition, or high humidity conditions

(c) Overcured laminates resulting from excessive additions of catalyst or accelerator, poor heat dissipation, high exothermic conditions

(d) Improperly resin-wetted laminates produced as a consequence of insufficient mechanical impregnation, incompatibility of sizing and fibers, or presence of resin-phobic contaminants

(e) Chemically degraded SPI standard control laminates in which the resin matrix, glass fibers, or both have been attacked by selective solvents, oxidizing agents, acids, or caustic compounds.

These and similarly prepared special samples were then examined under the stereomicroscope, metallographic optical microscope, and scanning electron microscope. Appropriate photographs were prepared of each type of failure or attack to serve as a set of "known condition" control laminates. Comparison with unknown exposure laminates ferrets out causes for their questionable or unsatisfactory performances. The examination of these test laminates is done under magnifications in the range of 100–13,000× and have proved extremely valuable in determining the resistivity of fiber reinforcements and the particular resin system towards the specific environment under consideration.

Infrared Spectroscopy. Identification of plastic materials of construction as to generic family and, wherever possible, as to commercially available resins or proprietary plastics is a necessary adjunct to plastic testing and plastographic analysis. It assists in determining the cause of poor performance in failure analysis studies, as an aid in establishing good standards, and promotes the design of better corrosion-resistant chemical processing equipment and structures. It is possible by infrared spectroscopy to use a very small segment of the plastic or resin material to identify rapidly and accurately specific polymers in the unit under examination.

Conclusion

It is necessary to use ingenuity in selecting or devising tests to determine the chemical resistivity of a particular plastic or resin exposed to a specific environment. The applications dictate the testing procedure needed to guarantee that the end results will give meaningful, safe, and useful engineering design data.

Literature Cited

1. Cass, R. A., Fenner, O. H., "Evaluating the Performance of Fiberglass Laminates," *Ind. Eng. Chem.* (1964) **56**, 29-34.
2. Fenner, O. H., "Plastographic Analysis Techniques in the Evaluation of FRP Structures and Equipment," *SAMPE Quart.* (1970) **1** (3) 26-31.

RECEIVED October 11, 1973.

20

Reinforced Plastics in Low Cost Housing

ARMAND G. WINFIELD and BARBARA L. WINFIELD

Armand G. Winfield Inc., 82 Dale St., West Babylon, N. Y. 11704

Our company designed and produced a low cost prototype house for Bangladesh to withstand cyclonic winds in excess of 150 mph and tidal waves. The house was constructed primarily of jute-reinforced polyester with a .010-inch exterior layer of glass fiber-reinforced polyester. The two-room house was 10 ft × 20 ft with an average height of 7.5 ft. Constructed of two monocoques, it has a dividing partition. The prototype successfully underwent full scale simulated cyclonic testing. A jet aircraft was used to create winds greater than 230 mph; 750 gal of water/min were released simultaneously to complete the cyclonic conditions. Surface temperatures of 200°F were recorded. Maximum deflection on prototype walls was 2.3 inches. The prototype indicates a major breakthrough in the use of plastics for low cost housing.

Early in 1972, CARE, Inc. initiated a program to establish goals for shelter using plastics materials for the Asian sub-continent, specifically Bangladesh, where a major war for independence had recently been fought. The war left tragic devastation and homelessness, and on the Island of Bhola, at the mouth of the Ganges, the effects of the war were magnified by a 1971 tidal wave which levelled this island and killed 225,000 persons in a single day. The Asian sub-continent—according to the United Nations—is one of the areas of the world critically in need of housing. This program is part of CARE's continuing effort to provide safe and inexpensive housing in the developing countries of the world.

Background

The most important low cost building system in use under CARE's aegis is the Cinvaram system. A composition of dirt, water, and 7–12%

concrete is prepared. A $50.00 Cinvaram press is operated by manual lever action and produces a "green" block. Two men can produce approximately 300 blocks per day, and the blocks must cure for 28 to 30 days prior to use. Cured blocks are placed on a peripheral concrete slab and are mortared together. Roofing made of corrugated galvanized iron sheets are attached to the Cinvaram by complicated roofing struts and trusses. A typical Cinvaram house 10 ft × 20 ft uses approximately 1500 blocks for walls and partitions.

A total of 7500 units were built in Bangladesh under CARE supervision in late 1972, and several thousand more were built in 1973. The Cinvaram structure has many drawbacks, including its inability to withstand hurricanes, tornadoes, earthquakes, or tsunamis. High winds can rip away the metal roofing, and earthquakes and floods can crumble or disintegrate its walls. To date, however, this seems to be the most ex-

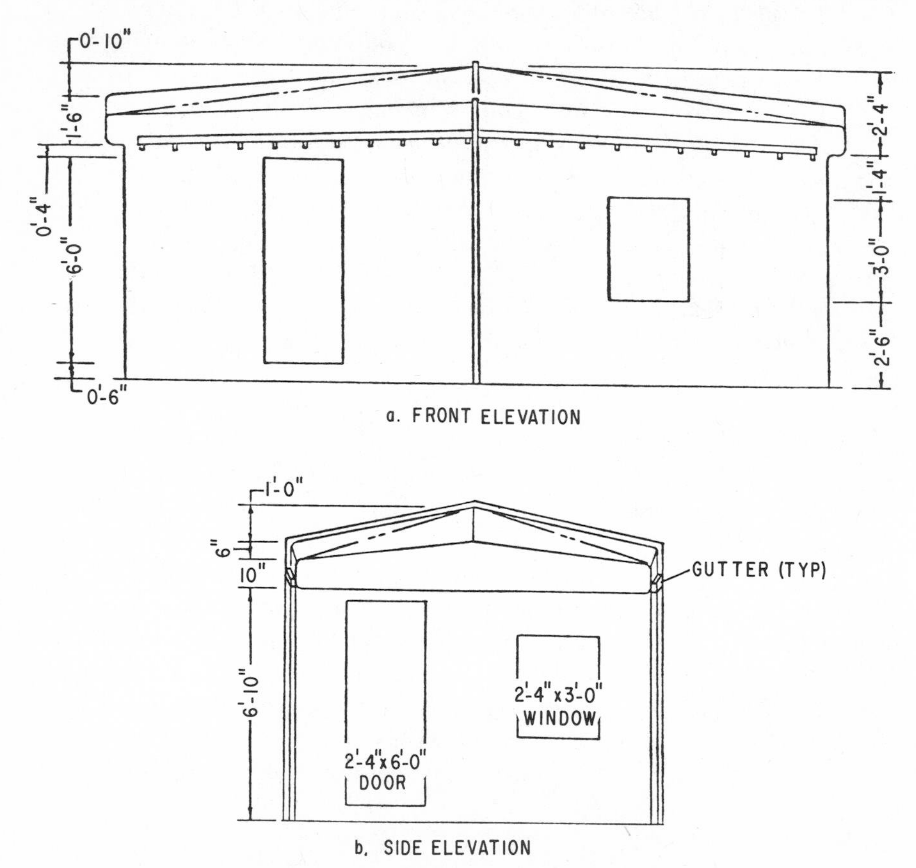

Figure 1. Initial house design by Anthony Marchese for Bangladesh as approved by CARE, Inc.

pedient, ethnically acceptable house for construction in many developing areas—one which replaces the open sky (or a tree) for a roof.

During the UNIDO meetings in Vienna, Austria, Sept. 20–24, 1971, entitled "Expert Group Meeting on the Use of Plastics in the Building Industry," one of the important conclusions was that plastics lend themselves best to mass production techniques which are required if the world's housing needs are to be met.

Although CARE personnel had experimented with plastics in housing in Bangladesh, prior to their war for independence, no successful structures had been produced. This work, however, created much local interest and indicated future possibilities.

Figure 2. Typical window construction (wood)

Research and Development

In early 1972, CARE, Inc. initiated a program with Armand G. Winfield Inc. for the material feasibility, design, prototype development, and testing of a low cost house for Bangladesh. Criteria for this house were to include:

(1) The use of jute—the main staple of Bangladesh—in association with a minimum of plastics materials.

(2) Ethnic and social acceptability to the Bangladesh people.

(3) Use of self-help in construction.

(4) Capability for high speed, on-site production by unskilled or semi-skilled local labor, in excess of 100 completed houses per day.

(5) Cost under $300.00 for each 10 ft × 20 ft structure.

(6) Ability to withstand cyclonic winds greater than 150 mph and other natural phenomena, including monsoon flooding and land upheavals.

After CARE briefings, initial designs were created by the architect, Anthony Marchese, and were approved by CARE, Inc. for prototyping (Figure 1). The house was to be 10 ft × 20 ft by an average height of

7.5 ft and was to be built as two monocoques, with an internal partition which would provide two rooms. The design included three doors—two external and one internal—and four windows, all of which opened inward (an ethnic necessity). The doors and windows would be wood—available on site in Bangladesh (Figure 2). Floors were not a consideration since custom dictated a tamped dung-earth floor. Bamboo verandahs, screens, and additional partitions would be constructed by the family once the basic house was erected.

Table I. Wind Tunnel Tests[a]

	Terrain Exposure		
	Center of large city	*Suburban area*	*Flat, open country*
Wind speed[a], mph	(1) 100 (2) 150	(1) 100 (2) 150	(1) 100 (2) 150
Design load[b], lbs/ft^2	(1) 15 (2) 34	(1) 29 (2) 65	(1) 48 (2) 106 (3/4 lb/in^2)

[a] Basic wind speeds based on air flow in open, level country at 30 ft above ground and for terrain exposure as noted.
[b] Design load = effective velocity pressure multiplied by appropriate pressure coefficient: 1.86.

A wind tunnel model was constructed from the approved design, and tests were conducted to prove its design and capabilities to withstand high wind loads (Table I). Engineering specifications were subsequently developed from these data and were used to select materials from which the prototype would ultimately be constructed. A foundation and anchoring system was developed by extending the walls of the house 18 inches below grade and simultaneously extending a 12-inch peripheral flange outward and perpendicular to the walls. This was further strengthened by including a net made of 2¼-inch width woven polypropylene tapes (450 lb test) which were interwoven on 12-inch centers in both directions of the house (Figure 3). The completed house was to be erected in an open pit 12 ft $\times$ 22 ft $\times$ 18 inches deep, and earth would be filled to ground level inside and outside the house. The weight of the earth on both the external flange and on the internal netting would hold the house firm during severe weather.

Various jute samples and products were collected in many forms—e.g., jute "heads," yarn, woven cloth, fibers, net, and rope. The initial goals for the house were based on the physical properties of glass-reinforced polyester. Using the criteria of 65–75% polyester resin in association with jute reinforcements, the minimum values would have to approach

18,000 psi flexural strength
850,000 psi flexural modulus
9,000 psi tensile strength.

First considerations dealt with combinations and ratios of jute fibers to polyester resins. Polyesters were chosen because they could be purchased economically for export, could be made self-extinguishing, and were compatible with the jute fibers. (Foamed polyurethane as a core material had failed in earlier beam tests.)

Jute in association with plastics materials was largely unexplored, except on a decorative basis, and little had been compiled on the physical characteristics of these combinations. The soft organic jute fibers, unlike inorganic glass fibers, absorb resin. Early experiments indicated that the jutes used provided little strength or stiffness in tensile and flexural tests. The addition of a glass fiber layer also added little strength to the laminate and in some combinations proved ineffectual or even detrimental as more resin soaked into the jute layers (Table II). Although the initial tests used jute, polyester resin (Polylite 94-158, Reichhold Chemicals, Inc.) and glass fibers had established the direction for material composites; the ratio between the jute and polyester remained to be developed further —as would the stiffness and strength.

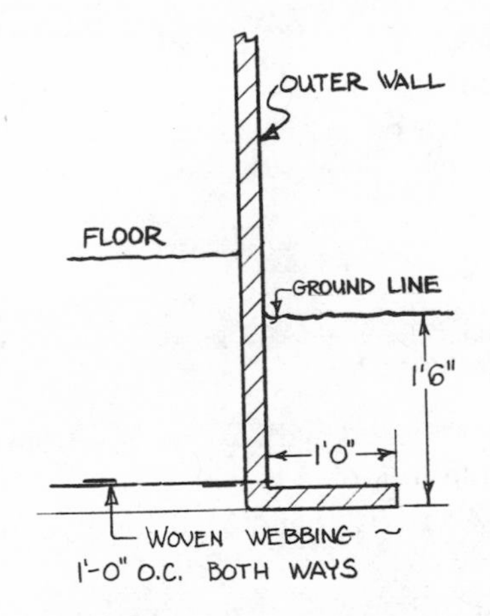

Figure 3. Anchoring system showing flange and netting

To improve the flexural strength and moduli, it was necessary to introduce finely divided inorganic fillers which would pass through the apertures of the airless spray-up equipment (Glas-Mate, Ransburg Florida, Inc.). Calcium carbonate filler (Atomite 319, Whitaker, Clark, and Daniels, Inc.), 20 wt % to the polyester resin, provided the acceptable properties. As shown in Table III, panel 1 A was made of four layers of jute saturated with unfilled polyester resin. The flexural moduli recorded were 444,000, 634,000 and 447,000 psi, respectively. Panel 1 B,

Table II. FRP Test Laminates[a]

	Flexural[b]			Tensile	
	Strength, psi	*Modulus, psi*	*Thickness, inch*	*Strength, psi*	*Thickness, inch*
Panel 1[c]*, resin-soaked jute*					
#1	11,700	232,000	0.037	2,600	0.032
#2	6,800	120,000	0.034	1,900	0.035
#3	6,400	63,000	0.034	—	—
Panel 15[d]*, 1 resin and glass only*[e]					
#1	23,000	674,000	0.054	6,600	0.050
#2	21,200	613,000	0.052	5,900	0.052
#3	18,800	555,000	0.052	—	—
Panel 16[d]*, 1 jute/1 resin and glass*[f]					
#1	21,500	576,000	0.066	4,000	0.072
#2	22,200	563,000	0.068	4,300	0.072
#3	9,000	510,000	0.067	—	—
Panel 13, 3 jute/2 resin and glass[g]					
#1	15,200	643,000	0.153	7,600	0.164
#2	15,000	577,000	0.157	10,000	0.161
#3	12,000	588,000	0.156	—	—
Panel 11, 4 jute/3 resin and glass[g]					
#1	11,700	548,000	0.238	8,200	0.260
#2	17,200	736,400	0.235	8,300	0.259
#3	13,900	642,300	0.235	—	—

[a] Standard conditions after 48-hr post cure at 150°F.
[b] Samples # 1 and 2 were tested with resin rich side in compression; sample # 3 was tested with glass side in compression.
[c] Jute only; note low physicals.
[d] Panels 15 and 16 were made simultaneously.
[e] Note higher physicals for glass alone.
[f] Note that addition of jute weakened the laminate.
[g] Note the more jute layers added, the weaker the composite became because the jute was absorbing resin and the glass became ineffectual.

which was made of four layers of jute staturated with 20 wt % Atomite to the polyester resin, shows flexural moduli of 549,000, 527,000, and 480,000 psi. Tensile strengths of panel 1 A are 5400 and 5100 psi *vs.* panel 1 B with readings of 7900 and 8200 psi.

Criteria for the polyester–jute composite were determined by two factors:

(1) It was necessary to use a filler to achieve stiffness without affecting other properties.

(2) Initially, woven jute cloths, nets, and tapes were absorbing resin in excessive amounts. Studies of the jute itself showed that its fibers are extremely porous and absorb resin. It was also determined that a double sheared and calendered weave of jute would minimize resin absorption. Thus, a 22 × 22 (thread count) Bengalon fabric was chosen as the fabric

for the prototype. (Bengalon is imported into the United States from India by White, Lamb, Finlay of Montclair, N.J.)

Ultimately, Panel 1 B (Table III) became the inner skin of the house composite. Since the foamed polyurethane core potential was abandoned because of core failures, a substitute core was necessary. A corrugation

Table III. FRP Test Laminates[a]

	Flexural[b]			*Tensile*	
	Strength, psi	*Modulus, psi*	*Thickness, inch*	*Strength, psi*	*Thickness, inch*
Panel 1A[c], *4 layers jute/unfilled resin*					
#1	10,700	444,000	0.0895	5,400	0.084
#2	12,100	634,000	0.079	5,100	0.088
#3	8,700	447,000	0.077	—	—
Panel 1B[d], *4 layers jute/with 20% Atomite in resin*					
#4	15,100	549,000	0.077	7,900	0.080
#5	12,500	527,000	0.0775	8,200	0.086
#6	11,700	480,000	0.079	—	—
Panel 1C[e], *2 layers jute/with 20% Atomite in resin*					
#7	11,800	54,200	0.0445	5,000	0.047
#8	12,900	65,200	0.044	5,600	0.046
#9	10,600	53,600	0.0455	—	—
Panel 1D[f], *1 layer jute/resin with 20% Atomite*					
#10	6,400	59,200	0.0375	4,100	0.028
#11	4,300	65,000	0.0465	3,700	0.040
#12	7,000	60,500	0.033	—	—
Panel 2A[g], *2 jute/unfilled resin/2 strands glass*					
#13	17,300	575,000	0.111	13,100	0.111
#14	18,800	573,000	0.118	10,500	0.104
#15	13,200	416,000	0.097	—	—
Panel 2B[g], *1 jute/resin with 20% Atomite/2 strands glass*					
#16	16,200	510,000	0.0855	7,800	0.094
#17	13,200	378,000	0.083	9,800	0.077
#18	13,700	422,000	0.077	—	—
Panel 2C[g], *2 jute/resin with 20% Atomite/2 strands glass*					
#19	13,200	553,000	0.092	12,900	0.087
#20	14,300	528,000	0.091	13,600	0.086
#21	11,500	616,000	0.084	—	—

[a] Tests done under standard conditions.
[b] Specimens # 1 and 2 were tested with resin-rich surface in compression; specimen # 3 was tested with glass rich surface in compression.
[c] Unfilled resin shows low flexural strength and modulus.
[d] Atomite increases flexural strength and modulus.
[e] Reduction of jute reduces flexural strength and modulus.
[f] Too weak.
[g] Used to determine criterion for outer skin.

of jute reinforced polyester proved satisfactory. Cements which could meet the physical requirements needed to fasten the core to the skins were developed and tested. The final formulation was made from rigid and flexible polyester resins stiffened with a thixotropic filler (Cab-o-sil, Cabot Corp.).

The final composite for the structure consisted of four layers of jute reinforced polyester for the inner skin; three layers of jute reinforced polyester for the corrugations; and one layer of jute reinforced polyester and one layer of .010-inch glass reinforced polyester for the outer skin (Figure 4). The exterior glass reinforced layer provided a barrier against weathering. The composite was an aerodynamic structure—*i.e.*, the skin could be pierced with a sharp blow, but the flexural moduli would preclude failure in high wind activity. Final beam tests on the completed composite structures proved that loads greater than 1800 lbs could be applied, with deflections up to .68 inch (Table IV).

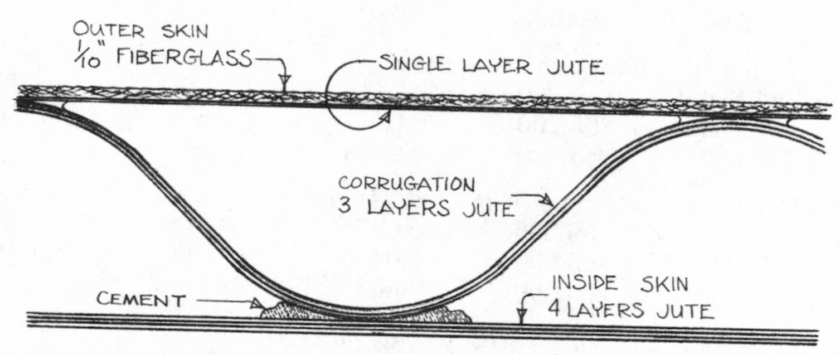

Figure 4. Final composite sandwich for prototype structure

Field Testing

Although laboratory testing had been conducted throughout the entire research and development phase, it was felt that full scale prototype testing would be more significant and conclusive than any single or sectional test. Further, the foundation/anchoring system could only be tested under cyclonic conditions. Grumman Aerospace Engineering Corp. provided its Calverton, L. I. test facilities for this purpose.

The fully assembled prototype house was transported from Armand G. Winfield Inc. to Calverton where it was anchored in the ground. The anchoring conditions were similar to those in Bangladesh (Figure 5). It was placed 50 ft from and parallel to a cross section of airport runway (Figure 6). Using a Navy A-6A Intruder jet fighter, the engine profile indicated that wind speeds approaching cyclonic forces could be gen-

Table IV. Flexure Test of Flat Sandwich Construction[a]

Panel

1–1 4-layer core, new cement[b], tested with woven roving side in compression
ultimate load—1810 lbs
ultimate deflection—0.65 inch
mode of failure—core failure

1–2 Same as 1–1, tested with woven roving in tension
ultimate load—1650 lbs
ultimate deflection—0.52 inch
mode of failure—compression in skin and core

2–1 3-layer core, tested with woven roving side in compression
ultimate load—1350 lbs
ultimate deflection—0.68 inch
mode of failure—core failure

2–2 Same as 2–1, tested with woven roving in tension
ultimate load—1050 lbs
ultimate deflection—0.51 inch
mode of failure—compression in skin and core

3–1 3-layer core, old cement[c], tested with woven roving in tension
ultimate load—730 lbs
ultimate deflection—0.63 inch
mode of failure—core failure

[a] Round steel bars, 2 inches in diameter, were used in loading. Specimens were loaded at two quarter-span points with the span length equal to 18 inches. The panel width was 6 inches. Load was applied at the rate of 0.10 inch/min, and defletion measured was crosshead movement.

[b] Ultimate formulation of rigid and flexible polyesters stiffened with a thixotropic filler (*see* text).

[c] Earlier, unsatisfactory cement formulation.

erated against the prototype at a distance of 50 ft. However, the surface temperature on the prototype skin at this distance would reach over 200°F. Anonometers would be used to clock the air speed of a single jet engine. To cool the prototype skin and to simulate cyclonic conditions further, 750 gal/min of water from a high speed pumper (fire fighting equipment) would be directed into the path of the jet.

Four tests were planned—each to last at least 5 minutes (Table V). Tests 5 and 6 were to be full jet blasts and would be run "to destruct." Tests 1, 2, and 5 would be aimed at wall No. 1—*i.e.*, the 10 × 10 ft end—while tests 3, 4, and 6 would be aimed at the 10 × 20 ft side.

Results

No previous jet tests were made to check out the location of the anonometer in relation to the jet blasts. In the first test, both jet engines were operating although only the use of one engine had been planned.

Figure 5. Completed prototype prior to testing at Grumman Aerospace Engineering Corp., June 1972

Inaccurate anonometer readings showed that this apparatus was between the jets rather than in line with the blasts. When the pilot accelerated both engines to get a reading, this blast ripped away the veranda, the tape grid, and external temperature gages. The pumper had not yet been turned on (Figure 7).

Extrapolation of the data from the spasmodic readings on the anonometer, coupled with the engine profiles as they reached peak acceleration, provided a conservative estimate (George Lubin and Peter Donahue, Grumman Aerospace Engineering Corp.) that by the time the water was

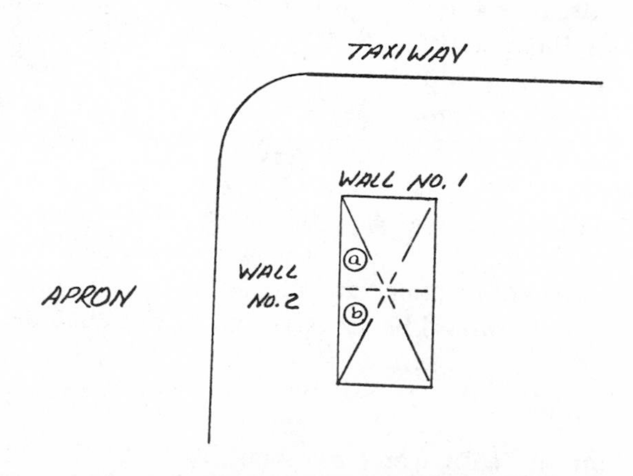

Figure 6. Ground plan of prototype adjacent to airport taxiways

Table V. Planned Test Sequence and Conditions

Test No.	*Wall No.*	*Wind Velocity, mph*	*Water Spray, gpm*
1	1	50	750
2	1	100	↑
3	2	50	|
4	2	100	|
5	1	136	↓
6	2	136	750

—Courtesy CARE, Inc.

Figure 7. Prototype during testing at Grumman Aerospace Corp., June 1972

released and the test completed, the force on the prototype was approximately 200 knots—*i.e.*, *ca.* 230 mph. The external skin on the end had buckled slightly from the heat (200°F) but relaxed after cooling. Tapes and gages were blown off and could not be used as conclusive evidence. The interior and exterior cameras were soaked by water which penetrated the wooden window shutters. Internal temperature rises were undetected. Although the prototype vibrated from this force, it did not move. The foundation held, and the house was left intact with only minor damage to its glass fiber reinforced polyester skin surface. The 10-ft wall span had deflected only 2.3 inches overall. Test No. 1 had, in effect, become tests 5 and 6 (Table VI). The other tests were anticlimactic and are covered in Table VI.

An additional series of tests on completed panels using a new joining system required for the second prototype was conducted at the DeBell and Richardson Testing Institute (Hazardville, Conn.) based on ASTM E-72. These tests showed deflection readings far in excess of the 2.3 inches of deflection noted in the spectacularly effective but limited Grumman tests.

Table VI. Revised Test Sequence and Conditions

Test No.	*Wall No.*	*Wind Velocity, mph*	*Water Spray, gpm*
1	1	220–230	750
2	2	40	↑
3	2	115	↓
4	2	170	750

Conclusions

Only by using inexpensive fillers as stiffeners (*i.e.*, calcium carbonate and jute) was this project successful. Jute as a replacement filler for glass fibers proved its effectiveness once the external fibers were removed by double shearing and calendering. Jute-reinforced polyester in such critical applications suggests many potentials in this direction but indicates the need for more research and development. The use of the materials described averaged $.4125/ft^2 of composite and brought the completed structure to paper costs within 15% of the $300 goal. The project from initiation through material feasibility studies, prototype design and construction, and full scale testing was accomplished in 16 weeks.

This development using plastics materials and inorganic fillers in association with indigenous native raw materials may result in the creation of safe, durable, and ethnically acceptable low cost housing for developing and developed countries.

Acknowledgments

The authors thank CARE, Inc., and particularly Louis Samia, Ralph Devone, Henry Sjaardema, Robert Cowan, and William Woudenberg for their cooperation. They also thank George Lubin for special technical assistance.

RECEIVED October 11, 1973.

INDEX

A

Abrasion resistance 134
ABS 23, 50, 63
  composites 4
Absorption 12
  oil 10
ACR 63
Acrylic modification of plasticized
    PVC 61
Additives 1
  powdered 106
Aggregate size, carbon black 174
Aging, humidity 139, 141
Alumina, hydrated 184
  -filled compounds 191
Alumina trihydrate 83
Amines 92
Aminofunctional silane in mineral-
    filled polyamide resins 75
Ammonium compounds, quaternary 91
Amosite 16, 30
Amylase, α- 167
Analysis techniques, plastographic 204
Analytical properties of conductive
    carbon blacks 175
Anthophyllite 16, 30
Antimony trioxide 184
  -chlorinated paraffin 185
Applications utilizing modifiers ... 67
Asbestos 116, 129, 157
  fiber 20
  filler 20
  mixtures, PVC/ 31
  in phenolics 26
  as a reinforcement and filler 16
  -reinforced rigid PVC 29
    glass *vs.* 39
Aspect ratio 2
Automotive applications 70

B

Bangladesh, low cost house for ... 209
Barium sulfate 91
Behavior of solids in bulk handling 106
Binder demand 10
Biodegradable fillers 160
  in thermoplastics 159
Bisphenol A–fumarate resin,
    propoxylated 199
Blends, dry 113
Blowing, extrusion 166
Bond failure, interfacial 2
Bond theory, reversible
    hydrolyzable 88
Boron/epoxy composites 138
Breaking strain 33
Brittle temperature 71, 182
Brodnyan's theory 99

C

Cab-o-sil 214
Calcium carbonate 91, 129
  fillers 10
  potential packing for 12
Carbon black 129
  aggregate size 174
  analytical properties of
    conductive 175
  -filled polymers 172
  grade and loading 175
  -loaded polymers, physical
    properties of 180
  in plastics, conductive 171
  porosity 174
  reinforcing filler 171
  -resin composites 2
  structure 174
Capillary rheometer 99
Catalytic effects in bonding resins
    to fillers 86
Chemical resistance of thermo-
    plastics 155
Chlorendic anhydride 199
Chlorinated paraffin, antimony
    trioxide– 185
Chopped glass process 125
Chrysotile 16, 30
Cinvaram system 207
Clay 91, 129
  silane treated 93
Coatings industry 7
Color changes in plastics 201
Composite(s) 1, 7, 137, 195, 207
  ABS 4
  asbestos 16, 29
  boron/epoxy 144
  boron-reinforced 138
  electrical properties of Novacite
    and Novakup in epoxy ...58, 59
  glass fiber reinforced 122
  graphite fiber-reinforced 138
  MBS 61
  mica 41
  modulus 42
  multidirectional 138
  nylon 5, 73
  polyester–jute 212
  polyphenylene sulfide 149

Composite(s) *(Continued)*
PVC 29, 128
quartz 52
stacking sequence, moisture as a function of 146
Compounder, twin-screw intermeshing and co-rotating 118
Compounding 164
of fillers 114
in motionless mixers 106
hot-melt 159
methods 115
polystyrene 167
Compounds, compression molding 155
Compounds, injection molding .. 150
Compression
molded PPS 156
molding 49, 108
compounds 155
Conductive carbon blacks in plastics 171, 175
Conductivity 172
effect of surface chemistry on .. 174
Corrosion engineering in reinforced plastics 195
Coupling agents 56
organofunctional silane 89
silane 55, 73
Crocidolite 16, 30
Crosslinking 182
Crystalline thermoplastics 4
Cure, resin 92
Cure time 91

D

Dapon diallyl phthalate resins ... 59
Deflectron test cell 203
Deformable layer theory 87
Degree of mixing 108
Differential thermal analysis scan .. 188
Diisodecyl adipate 67
DOP 65
Dry blends 113

E

Effect of increasing temperature on resistivity 177
Effect of moisture 137
Effect of surface chemistry on conductivity 174
Einstein-Guth-Gold equation 2
Elastic effects 99
Elastomers 171
Electrical properties of Novacite and Novakup in epoxy composites 58, 59
Electrical tapes, modifiers in 68
Electron microscopy, scanning ... 159
Elongation 181
Environment, incompatibility of a reinforced plastic with 159, 201
Enzyme attack 167
Epichlorohydrin 198
Epoxy 198
composites, boron/ 144
composites, electrical properties of Novacite and Novakup in 58, 59
exotherms 92
resin 138
Ethylene–vinyl acetate 180
Exotherms
epoxy 92
maximum 89
polyester 90
resin 91
Extenders 7, 53
Extensibility 132
Extruders, single-screw 121
Extrusion blowing 159, 166
Extrusion-blown LDPE film 168

F

Failures in glass fiber reinforced plastics 197
Feldspar 188
Fiber(s)
asbestos 20
distribution in plastics 38
glass reinforced plastics 3
graphite 138
jute 207
length distributions 100
rheology of suspensions of 95
Fibrillation 167
Filled polyphenylene sulfide compositions 149
Filler(s) 1, 7, 196
asbestos 20
as a reinforcement and 16
biodegradable 159
calcium carbonate 10
carbon black 171
compounding of 106, 114
in motionless mixers 106
fire-retardant 184
inexpensive 218
inorganic 211
interfaces, nylon/mineral 76
loading 111
non-reinforcing 116
in paint 8
polytetrafluoroethylene 5
and reinforcements 1
reinforcement of plasticized PVC 128
–resin interface, mineral 84
selection, biodegradable 160
settling properties 192
silane-treated 86
silicas as 2
in thermoplastics, biodegradable 159
types of 116
Film, extrusion-blown 166
Film, polyolefin 159
Fire-retardant filler 184
Flammability 38
Flattening agents 8

Flex, quasi-isotropic 143
Flexural creep and modulus 151, 153
Flexural strength ...25, 48, 81, 203, 211
unidirectional 143
Flexural testing 140, 215
Flow behavior 112
Fold endurance 69
Fumarate resin, propoxylated
bisphenol A– 199
Furan 199
Fybex 79, 116, 154

G

Glass 129
vs. asbestos-reinforced rigid PVC 39
fiber 157
reinforced composites 122
reinforced plastics 197, 200
reinforced thermoplastic
polymers 120
fibers in polypropylene,
suspensions of 98
fibers in silicone oil 99
process, chopped 125
microbeads 91
reinforced plastics, fibrous 3
reinforced polymers 153
silane-treated 91
Graphite fiber-reinforced
composites 138

H

Hardness 37, 129, 201
of quartz 53
Shore A 68
Heat deflection temperature 25, 36
Heat distortion temperatures 141
Heat stability 20
procedure 164
Histograms 164
Hot-melt compounding 159
Hot strength 134
House for Bangladesh, low cost .. 209
Housing, reinforced plastics in low
cost 207
Humidity aging 139, 141
Hydrated alumina 184
-filled compounds 191
Hydrolyzable bond theory,
reversible 88

I

Impact modifiers 63
Impact strength 33, 84
Incompatibility of a reinforced plas-
tic with the environment ..159, 201
Inexpensive fillers 218
Infrared spectroscopy 205
Injection molding 71
compounds 47, 150
Inorganic fillers 211
Interface, mineral filler-resin 84
Interface, nylon/mineral filler 76
Interfacial bond failure 2
Isophthalates 199
Izod impact tests 111

J

Jute 210
composite, polyester– 212, 218

K

Kaolin clay 80, 99

L

Laminal wall attack 202
Laminate 197
LDPE 164
film, extrusion-blown 168
Light scattering, narrow angle ... 159
Loading, carbon black grade and 175
Lorite 14
Low cost house for Bangladesh .. 209
Low quartz microforms 52
Low-temperature flexibility 131

M

Matrix, mica in a polycarbonate .. 45
Maximum exotherms 89
Maximum packing 13
MBS 63
Mica in a polycarbonate matrix .. 45
Mica-reinforced thermoplastics ...50, 51
Microform technology 54
Micromechanics 2
Microscopy, scanning electron ... 159
Mineral-filled nylon 73
Mineral filler 94
interface, nylon– 76
–resin interface 84
Mining 57
Mixing, degree of 108
Mixing, simple 107
Modifiers, applications utilizing .. 67
Modifiers, impact 63
Modifiers in electrical tapes 68
Modulus 25, 129, 211
composite 42
flexural 203
Young's 110
Moisture, effects of 137
Moisture as a function of composite
stacking sequence 146
Molding, compression 49
Molding, injection 71
Molecular weight 39
distribution 39
Mooney equation 2
Motionless mixers, compounding of
fillers in 106
Multidirectional composites 138

N

Narrow angle light scattering 159
Non-reinforcing fillers 116
Notched izod impact strength 48

Novacite and Novakup in epoxy composites, electrical properties of 58, 59
Novakup and Novacite polypropylene 56
Novaculite silica 81
Nulok 80
Nylon 51, 55
  composites 5
  mineral-filled 73
  /mineral filler interfaces 76
  as the resin component 76

O

Optimizing cost/performance properties 78
Organofunctional silane coupling agents 89
Oxygen index 191
  test 185

P

Packing 10
  maximum 13
Paint, fillers in 8
Paperbestos 30
Paraffin, antimony trioxide–chlorinated 185
Particle size distribution 11
Percent elongation 110
Phenolics 24, 199
  asbestos in 26
Pigment volume concentration, critical 10
Pigmentation 196
Plastibest 30
Plastic(s)
  chemistry, fillers and reinforcements in 1
  conductive carbon blacks in 171
  consumption of reinforced 3
  with the environment, incompatibility of a reinforced 201
  glass fiber reinforced 3
    failures in 197
    testing of 200
  in low cost housing, reinforced .. 207
  rheology of reinforced 98
Plastographic analysis techniques .. 204
Plasticized PVC 61, 128
Plasticizers, polyester 65
Plasticizing agent, water as a 148
Platelet reinforcement 41
Polycarbonate matrix, mica in a .. 45
Polyester(s) 199
  casting compounds, styrene– .. 184
  exotherms 90
  jute-reinforced 218
  plasticizers 65
  resins 86
Polyethylene 23
  –vinyl acetate 171
  /starch sheet 167
Polyimide 142
  reinforced 4
Polylite 211
Polymer(s)
  carbon black-filled 172
  glass fiber reinforced thermoplastic 120
  glass-reinforced 153
  physical properties of carbon black-loaded 180
  /stabilizer combination 39
  thermoplastic 171
Polyolefin film 159
Polyphenylene sulfide (PPS) compositions 149
  compression molded 156
Polypropylene 20, 123, 166
  resin 107
  suspensions of glass fibers in .. 98
  talc-filled 110
Polystyrene 24
  compounding 167
Polytetrafluoroethylene filler 5
Poly(vinyl chloride), *see* PVC
Porosity, carbon black 174
Porosity index 13
Potassium titanate microfibers ... 4
Potato starch grains 163
Potential packing for calcium carbonate 12
Powdered additives 106
Processing characteristics 112
Properties, filler settling 192
Properties, unidirectional 138
Propoxylated bisphenol A–fumarate resin 199
PVC 20
  acrylic modification of plasticized 61
  /asbestos mixtures 31
  glass *vs.* asbestos-reinforced rigid 39
  rigid 61
  semi-rigid 67

Q

Quartz, hardness of 53
Quartz microforms, low 52
Quasi-isotropic flex 143
Quasi-isotropic tension 143
Quaternary ammonium compounds 91

R

Raw starch 159
Reinforced plastics
  corrosion engineering in 195
  with the environment, incompatibility of 201
  in low cost housing 207
  rheology of 98
Reinforced polyimides 4
Reinforced PPS 152
Reinforced thermoplastics 4
Reinforcement
  asbestos as a 16

Reinforcement (*Continued*)
carbon black as a ... 171
in plastics chemistry, fillers and ... 1
of plasticized PVC ... 128
platelet ... 41
Residence time ... 116
distribution ... 116
Resilience, rubbery ... 133
Resin(s)
aminofunctional silane in mineral-filled polyamide ... 75
composites, carbon black ... 2
composites, high performance structural ... 137
component, nylon 6 or nylon 6,6 ... 76
cure ... 92
Dapon diallyl phthalate ... 59
exotherms ... 91
interface, mineral filler– ... 84
polyester ... 86
polypropylene ... 107
propoxylated bisphenol A–fumarate ... 199
thermosetting ... 86
Resistivity, effect of increasing temperature on ... 177
Response surface ... 47
Restrained layer theory ... 86
Reversible hydrolyzable bond theory ... 88
Rheometer, capillary ... 99
Rheology, fiber suspension ... 99
Rheology of reinforced plastics ... 98
Rice starch grains ... 163
Rigid PVC ... 61
asbestos-reinforced ... 29
Rod climbing effect, Weissenberg ... 102
Roving process ... 123
Rubbery resilience ... 133

S

SAN ... 51
Scanning electron microscopy ... 159
Scattering, narrow angle light ... 159
Selection, biodegradable filler ... 160
Semi-rigid PVC ... 67
Shear strengths ... 139
Shore A hardness ... 68
Shrinkage ... 201
Silane ... 92
aminofunctional ... 75
coupling agent ... 55, 73
organofunctional ... 89
-treated clay ... 93
-treated fillers ... 86
-treated glass ... 91
-treated zircon ... 86
Silanois ... 56
Silica ... 91, 129, 188
novaculite ... 81
Silicas as fillers ... 2
Silicone oil, glass fibers in ... 99
Silicosis ... 194
Single-screw extruders ... 120
SiO_4 tetrahedra ... 53
Solids in bulk handling, behavior of ... 106
Sparking ... 197
Spectroscopy, infrared ... 205
Spherical particles ... 97
Spherulite size measurement ... 164
Soil burial tests ... 159
Starch grains, potato ... 163
Starch grains, rice ... 163
Starch sheet, polyethylene/ ... 167
Starch, raw ... 159
Strength ... 110
flexural ... 48, 203, 211
hot ... 134
notched izod impact ... 48
shear ... 139
tensile ... 66, 81, 132, 142, 151, 180
unidirectional flex ... 143
yield ... 181
Structure ... 174
Styrene–polyester casting compounds ... 184
Surface changes ... 201
Surface chemistry on conductivity, effect of ... 174
Surface functionality ... 56
Surface, response ... 47
Suspensions of fibers, rheology of concentrated ... 95
Suspensions of glass fibers in polypropylene ... 98
Swelling ... 201

T

Talc ... 91
-filled polypropylene ... 110
Technology, microform ... 54
Tension, quasi-isotropic ... 143
Temperature
brittle ... 71, 182
heat deflection ... 25, 36
heat distortion ... 141
properties, low ... 182
on resistivity, effect of increasing 177
Tensile
modulus ... 32
properties ... 70
strength ... 33, 44, 66, 132, 142, 151, 180
flexural ... 81
Test, oxygen index ... 185
Testing of glass fiber-reinforced plastics ... 200
Tests, soil burial ... 159
Tetrachlorophthalic anhydride ... 199
Thermoforming ... 167
Theory, deformable layer ... 87
Theory, restrained layer ... 86
Theory, reversible hydrolyzable bond ... 88
Thermal stability ... 149
Thermoplastic polymers ... 120, 171

Thermoplastics 166
biodegradable fillers in 159
chemical resistance of 155
crystalline 4
reinforced 4
Thermosets, mica-reinforced 50, 51
Thermosetting resins 86
Titanium dioxide 9
"Titration" technique 164
Transfermix system 121
Twin-screw extruders 121
Types of fillers 116

U

Ultimate elongation 67
Unidirectional flex strength 143
Unidirectional properties 138

V

Vinyl acetate, ethylene– 180
Viscosity 7, 96, 155, 161, 192

W

Water boil exposures 139
Water as a plasticizing agent 148
Weight changes 202
Weissenberg rod climbing effect .. 102
Wind tunnel model 210
Wollastonite 77, 99
–Fybex 79
Wood flour 3, 161

Y

Yield strength 181
Young's modulus 110

Z

Zircon 91
silane-treated 86

The text of this book is set in 10 point Caledonia with two points of leading. The chapter numerals are set in 30 point Garamond; the chapter titles are set in 18 point Garamond Bold.

The book is printed offset on Danforth 550 Machine Blue White text, 50-pound. The cover is Joanna Book Binding blue linen.

Jacket design by Linda McKnight.
Editing and production by Mary Westerfeld.

The book was composed by the Mills-Frizell-Evans Co., Baltimore, Md., printed by The Maple Press Co., York, Pa., and bound by Complete Books Co., Philadelphia, Pa.